W9-CAP-368

NEMESIS

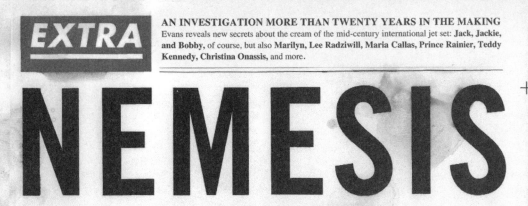

AN INVESTIGATION MORE THAN TWENTY YEARS IN THE MAKING
Evans reveals new secrets about the cream of the mid-century international jet set: **Jack, Jackie, and Bobby**, of course, but also **Marilyn, Lee Radziwill, Maria Callas, Prince Rainier, Teddy Kennedy, Christina Onassis**, and more.

NEMESIS

USA $25.95 / Canada $39.95   ## THE TRUE STORY   *VOLUME ONE* 2004

Aristotle Onassis, Jackie O, and the Love Triangle That Brought Down the Kennedys

AN **EXTRAORDINARY STORY** of three of the world's most powerful men and the femme fatale who came between them. Peter Evans has been researching the life and times of Aristotle Onassis for more than two decades; he interviewed not only Onassis, but scores of his intimates. At the core of his reporting—and of this riveting book—is the story of the longtime rivalry between Onassis and the Kennedy clan.

CONTINUED INSIDE

"As an investigative reporter, Peter Evans must be counted among the very best." — *Eric Clark*

ReganBooks
Celebrating Ten Bestselling Years
An Imprint of HarperCollinsPublishers

Peter Evans

ISBN 0060580534

HarperCollins books may be purchased for educational, business, or sales promotional use. For information please write: Special Markets Department, HarperCollins Publishers Inc., 10 East 53rd Street, New York, NY 10022.

FIRST EDITION

Designed by Kris Tobiassen

Printed on acid-free paper

Library of Congress Cataloging-in-Publication Data has been applied for.

ISBN 0-06-058053-4

04 05 06 07 08 RRD/DIX 10 9 8 7 6 5 4 3 2 1

AND THIS ONE IS FOR MARK AND CHRISTINE

CONTENTS

INTRODUCTION

"When he hated nobody was spared. . . ."

—CHRISTINA ONASSIS

I first met Aristotle Onassis on January 5, 1968, at his apartment on avenue Foch, Paris. Thinking of "making a book" about his life, he had written to his friend Jean Paul Getty asking him to recommend a writer. Probably because I had recently spent some time interviewing him for a magazine profile, Getty had simply returned the letter, with my name and telephone number written in the margin. It was good enough for Onassis—was I interested?

Onassis was probably the most famous millionaire in the world; his affair with Maria Callas had made him sexy as well as notorious. Of course I was interested—but not if he just wanted a hagiographer. I asked what kind of book he had in mind. He started to tell me the story of his life: his childhood in Smyrna, now Izmir: how he had survived the massacre of 1922 and became the head of his family when his father was thrown into prison by the Turks, their escape to Athens, and his flight to Argentina, where he made his first fortune.

Although I knew that I was seeing a performance, the alchemy of money, sex, and mystery that was at the core of his story was irresistible. Over lunch at Maxim's, I took the bait.

I quickly discovered that he did not respond well to question-and-answer interview techniques. He was at his best when he was free-associating—preferably over a meal, in a bar, or simply walking the streets late at night in London, Paris, or New York. Like all spellbinders, he liked to have an audience. In Paris, Johnny Meyer, an old Hollywood publicity man who described himself as Onassis's *aide-de-camp,* sometimes tagged along, with an elegant companion from Madame Claude's callgirl establishment or a showgirl from the Crazy Horse. Occasionally, Onassis's son, Alexander, joined us, but he seemed ill at ease listening to Onassis's stories. Theirs was not a happy relationship, although Alexander, a kind and decent young man, was always respectful to his father. The man who inspired Onassis best was Constantine Gratsos, his oldest and closest friend, and the architect of many of his biggest schemes. "I will sometimes lie to you," Onassis once told me. "But Costa will always tell you the truth."

Although he had great charm, and could be extraordinarily generous, he was also capable of sadism toward those closest to him, and his mood could switch at alarming speed: Exhilaration could turn to despair, and not only when he drank. When I mentioned Stavros Niarchos's name—before I knew enough to tread warily around anything to do with his rival, brother-in-law, and the man who was soon to marry his ex-wife and the mother of his children, Tina Livanos—he left the room, slamming the door so hard that I expected never to see him again. Nevertheless, he continued to come and go through the spring and summer months of 1968. I would receive a call to meet him in Paris, where we would sometimes talk for many hours, often through the night. Sometimes, though, he granted me only a few minutes, and once he failed to find time to talk to me at all after what had seemed to be a most urgent summons. I worked with him in this fashion for ten months before he canceled the deal; a week later, on October 20, 1968, he married Jacqueline Kennedy.

In the spring of 1974, Onassis told me that he wanted to resume work on the book. His marriage to Jackie had been a sham. On paper, it had lasted six years; in reality, it was over within weeks of the nuptials on his private island of Skorpios—marked from beginning to end by calamity. Alexander had been killed. Tina had died. His daughter Christina had been married and divorced, and her life was a mess. He had lost Monaco, the principality over which he ruled for more than a decade, and his dream of making a new fortune in partnership with the colonels' dictatorship in Greece had collapsed. An even more ambitious plan to take over Haiti had ended badly.

He was suffering from myasthenia gravis, a disorder of the body's autoimmune system, which made it difficult for him to talk. We both knew that he was dying, and when I asked him why he had become interested in the book again, he told me, "I hate leaving anything unfinished."

But he was a curious as well as a vain man, and I knew that what I would write about him when he was dead mattered to him. Throughout the next six or seven months, until his death, in March 1975, we continued to work on the book we both knew he would not live to see published. Meanwhile, his instinct for self-preservation had degenerated with age and illness into paranoia (obsessed by tricks and games to test the loyalty of his family and friends, he never lost the power to unnerve), and our book was his last attempt to fix how the world would remember him.

He clearly believed he had achieved his purpose the last time we said goodbye. "*What you don't know now,*" he told me with the slow, polite drawl that had deceived lawyers, financiers, and lovers for years, "*nobody will ever know.*"

Six weeks later Aristotle Socrates Onassis was dead.

Shortly after it was published in 1986, I took a copy of *Ari: The Life and Times of Aristotle Socrates Onassis*[1] to Yannis Georgakis in Athens. A distinguished Greek lawyer, former chairman of Olympic Airways, and Onassis's house intellectual and *raisonneur,* Georgakis had joined the Athens University Law School at fifteen and graduated at twenty. He continued to pursue his studies at the universities of Munich, Heidelberg, and Leipzig, where he obtained his doctorate in criminal law in 1938. He taught there until he returned to Greece at the outbreak of World War II.

Georgakis had been my most reliable informant and a wise sounding board after Onassis's death. A discreet homosexual with a gentle manner and disarming sense of humor, he was not, however, someone to be trifled with; during the German occupation, he defended bravely and without remuneration hundreds of Greek resistance fighters before German courts martial. He regarded Onassis with affection, yet he could be tough on him, too: Onassis, he had warned me the first time we met, was "a charming psychopath" bound by "absolutely no moral imperatives at all." Unless I understood that, he said, I would never understand anything about Onassis at all.

Although he had become engrossed in his new role as Greek ambassador-at-large to the Arab world, he had read each draft of my book, and checked the galleys, with a lawyer's attention to detail; the first edition I took to Athens was a token of my gratitude. I believed I had written the last word on the subject of Aristotle Socrates Onassis. I was disappointed when a few days later Georgakis told me that although I had gotten closer than any writer ever would to the truth about Onassis, I had also "missed the real story."

It was not what I had expected or wanted to hear. I should have looked more closely at the events surrounding Onassis's marriage to Jackie Kennedy, he said. He liked to tease me with ambiguity, but I was in no mood for games. Having arranged the wedding on Skorpios, he was uniquely qualified to know the truth—why didn't he just tell me what I'd missed?

"I can't. It would destroy Christina," he answered, defensively. "Let sleeping dogs lie."

I knew that his position as permanent secretary of the international committee of the Alexander Onassis Foundation put him in a delicate situation. I tried to refine my questions, to draw him out without embarrassing him. But he was too smart a lawyer not to see through that.

In the following months, as information trickled in from unexpected sources, as it always does after publication of any biography, hints of a more complex story began to emerge.

*   *   *

"A point is the beginning of all geometry; a percentage point is the beginning of all fortunes. Before that point nothing matters," Onassis had told Christina when she asked about his fortune-making years. I knew that she had never read *Ari*. "I'm afraid of what I might discover," she told me, although in truth she had never shown much curiosity about her family's past.

In May 1988, I had lunch with her in Paris and gave her a videotape of *The Richest Man in the World,* a Hollywood miniseries of my biography. I hadn't seen her as often as when her father was alive and, at thirty-seven, she had changed in ways I still find difficult to define. "I used to think that being rich was some kind of nirvana. It isn't. I've discovered that since my father died," she told me. I asked whether that had made her harder or more vulnerable. "Wiser," she answered.

I heard nothing from her for five months. At the beginning of October, she called me in London. She was shortly going to Buenos Aires to celebrate a friend's fortieth birthday and suggested that we have lunch in Paris before she left. She had finally read *Ari* and wanted to discuss it with me, she said.

We lunched at a cafe on rue Capucine, behind the place Vendrome. She had shed about forty pounds in a Swiss clinic since our last meeting, but was still heavier than in 1970, when I had first taken her out in London; even after four failed marriages and numerous unhappy love affairs, her face was still gentler than it looked in press photographs. She had just returned from Geneva, where she had rewritten her last will and testament for the eighth time since her father died. Writing her will always gave her pleasure. "If you want to find out how you really feel about somebody, write your will. It concentrates the mind wonderfully," she told me mischievously.

She talked about the men who were not remembered in the will, men who had let her down and used her. No wonder my head is so messed up, she said, although I knew that her heavy prescription-drug habit didn't help her state of mind much either. She took risks her Parisian friend and confidante Florence Grinda begged her not to take. After her first suicide attempt in 1974, and ever since her uncle, Dr. Theodore

Garofalides, had prescribed the antidepressant drug imipramine, she had been on an upward curve of amphetamine addiction. Unable to use a needle on herself, she had hired a permanent private nurse to administer the shots.

Christina talked about Thierry Roussel, her last husband and the father of her daughter, Athina, who was a few months short of her fourth birthday. She knew that Roussel, whom she still loved, was not liked by her friends and was distrusted by her advisers, who were appalled at the way he had two-timed her with his Swedish mistress. "Thierry is a Frenchman," she shrugged, as if that explained everything.

I suspected that she was not telling me these things without a purpose. But was it to prove that she trusted me or to appear unguarded about one thing in order to dupe me about something else? The only thing I could be sure of was that she was her father's daughter, despite the plastic surgery on her nose and jaw line, she still had his looks about the eyes and mouth, and quite a bit of his charm and wiles, too. I was still trying to figure out her motives when she told me that she liked my biography of her father. I'd gotten him almost right most of the time, although I had failed to understand his capacity to hate.

"When he hated, nobody was spared. It destroyed us all in the end," she said. It was an extraordinary statement, and I was still wondering how to respond to it when she added: "But he was not a bad man. He did some bad things, he made some bad mistakes, but he was never *deliberately* wicked."

Onassis had been dead thirteen years, but talking about him touched her to the point of tears. Although I knew that her relationship with her father defied her comprehension, I was moved by her loyalty and her determination to defend his name. Although she was having difficulty explaining exactly what troubled her, I gathered that she wanted me to write a magazine piece about Onassis's marriage to Jackie Kennedy that would clarify some issues I hadn't dealt with in my book. Georgakis had told me I had missed the real story, and I wondered whether this was the same lacuna that was troubling Christina. I'd planned to fly back to London that evening, but having seen her in any number of emotional states in the past, I knew that she would not be able to continue our conversation coherently that night, and I decided to stay over.

The following day, we met at a cafe on the Quai des Grands-Augustins. Her energies revived by a good night's sleep, she seemed much stronger and more purposeful. Having made me swear to keep her confidence, and not to use the story without her agreement, she told me—like a lawyer summarizing the evidence for the defense—what was causing her so much consternation.

In 1968, she told me, Onassis had paid a Palestinian terrorist named Mahmoud Hamshari (who later became a leading figure in Black September, the terror group responsible for the massacre of eleven Israeli athletes at the 1972 Munich Olympics) protection money for his airline, Olympic Airways. Later, Onassis learned that his money had been used to finance the assassination of Robert Kennedy.

Even from this brief outline her concern was understandable. For Bobby Kennedy's death had cleared the way for Onassis to marry his brother John F. Kennedy's widow—a marriage that Bobby had sworn would happen only over his dead body. Since nothing was allowed to be accidental around the Kennedys—theories grew around the family like Boston ivy—Christina feared that if the link were discovered, if somebody followed the money, who would believe that her father had not been directly responsible for setting up the murder?

I had known Christina for twenty years, and although I felt that there was something she was not telling me, I knew this was not the time to press her. We agreed to meet again, together with Georgakis, at the beginning of December. "Perhaps you could come over for my birthday," she suggested. She would be thirty-eight on December 11 and had arranged a party at Maxim's. I wouldn't miss it for anything, I told her.

Three weeks later she was found dead in her bath in Buenos Aires, the city where her father had made his first fortune sixty years before.

Georgakis refused to discuss the matter with me after Christina's death, but when I told him that I had been asked to write her biography, he told me, "Do it if you want to, but she's a sideshow. Her father's still the story."

It was more than three years before he finally agreed to help me unravel the secret that was at the heart of Onassis's marriage to Jacqueline Kennedy—and the truth about the murder of Bobby Kennedy.

"It isn't quite as straightforward as Christina thought it was," he began, with a sense of understatement that I now realize perfectly defined the man.

I must, he said, begin with the premise that, for Onassis, Bobby Kennedy was unfinished business from way back. . . .

ONE

THE BLOOD TRADE

**Were one to ask me in which direction
I think man strongest,
I should say, his capacity to hate.**

—H. W. BEECHER, 1813–1884

Robert Kennedy and Aristotle Onassis met for the first time at a cocktail party given by the English socialite Pamela Churchill* at the Plaza Hotel in New York City in the spring of 1953—the year Jacqueline Lee Bouvier married John F. Kennedy.

Pamela Churchill was a shrewd networker long before the term had been invented, and her guest list had been drawn from the elite of the American establishment and the world's richest people. Daughter of an English baron, and the former wife of Randolph Churchill—the drunkard son of the British prime minister—Pamela, who would become the model for the elegant tramp Lady Ina Coolbirth in Truman Capote's *Answered Prayers*,[1] knew the great and near great of five continents. It was said that for legendary amounts of money, she had slept with many of them.

---

* Later Pamela Harriman, U.S. ambassador to France.

She had known Bobby since 1938, when his father, Joseph P. Kennedy, was the American ambassador to England. She and Bobby's older sister Kathleen were debutantes together in the last London season before the start of World War II, and had remained friends until Kathleen's death in a plane crash in 1948.

Onassis was not such an old friend. Since Pamela's ex-husband Randolph had introduced them in the South of France several months earlier, however—an introduction that Onassis said had cost him £2,000 (some £40,000 in today's currency)—the Greek shipping millionaire had become a close one (her lover, he said; not so, she protested, although her veracity in such matters was as questionable as Onassis's). Onassis was far too earthy for her tastes, Pamela told friends.[2] An unmistakeable *arriviste,* he possessed a volatile temper, especially when he'd had too much to drink, and his habit of smashing plates and making scenes in restaurants offended her English sensibilities.

Although Onassis was attracted to Pamela's world, and knew he would be accepted more easily if he adopted the elegant dress, language, and manners of their class—much as his brother-in-law, Stavros Niarchos had done—he refused. "I won't play the hypocrite for anyone," he told his young, English-educated wife Tina, daughter of the 1930s shipping king Stavros Livanos, when she tried to break him out of his Greek chrysalis and repackage him as an English toff.[3]

Nevertheless, Pamela Churchill was a practical woman, and it was clear that her interest in Onassis had been rekindled—and her sense of tolerance restored—by the news that he had just bought the principality of Monaco. More precisely, hiding behind a maze of Panamanian fronts, he had acquired SBM,* a moribund property company that owned an Edwardian pile of real estate in Monte Carlo, including the casino, the yacht club, the Hotel de Paris, and about one third of the principality's 375 acres.

Situated between the oil fields of the Middle East and the markets of Europe and North America, Monte Carlo was a perfect base for Onassis's

---

* Society des Bains de Mer et Cercle des Etrangers.

operations. The climate pleased him, the social life met with Tina's approval, and the principality was tax free.

Overnight, Onassis had become famous; suddenly, everything he did was news. His wealth, as well as the hints of something undisclosed about his past, made wonderful copy. More than just another rich Greek, this small, dark, sybaritic figure with sensual heavy-lidded eyes was recognized in the street. Women began to proposition him as if he were a movie star; he took to wearing dark glasses and engaged a public relations man. Reporters dubbed him the "king of Monaco" (a tabloid ennoblement that did not go down well with Rainier, the prince of Monaco). He gave interviews on how to handle women: "I approach every woman as a potential mistress," he said. "Beautiful women cannot bear moderation; they need an inexhaustible supply of excess."

But his love affair with the media was not entirely motivated by ego. His cultivated image as a mysterious but magnanimous rags-to-riches tycoon also "sanctioned his sharp deals," in the words of one American aide.[4] And no deal had been sharper than his acquisition, five years earlier, of ten U.S. surplus T2 tankers. Because of their size and strategic significance, the ships had been forbidden to foreigners, but at $1.5 million each, they had been an irresistible purchase for Onassis, and with the help of a U.S. corporation fronted by three American citizens, he easily circumvented the exclusion clause and bought them.*

---

* United States Petroleum Carriers, Inc., was set up with an authorized capital stock of 1,000 shares. Six hundred shares of this stock were issued to three Americans: Robert L. Berenson, Robert W. Dudley, and Admiral H. L. Bowen. Four hundred shares remained unissued. On December 30, 1947, the Maritime Commission approved the sale of four T2 tankers to United States Petroleum Carriers, Inc. Almost immediately, Dudley bought out the admiral's interest, and one week later Berenson acquired Dudley's shares for $125,000 (original cost $1,000). On the same day, the Sociedad Industrial Maritime Financiera Ariona, Panama, S.A., owned by Onassis and his cousins Nicholas and Constantine Konialidis, purchased the remaining 400 shares of the authorized stock. In the next six months, the same Panamanian corporation bought an additional 90 shares of stock from Berenson, giving them a 49 percent interest in the corporation. At the same time, Berenson reduced his own personal holdings to 48 percent by selling 10 shares to each of three American citizens closely tied to Onassis (Clif-

*   *   *

Robert Kennedy also became front-page news for the first time in 1953. And if the 1950s were not to be *his* glory years, as they were for Onassis, they were unquestionably heady ones.

Small and more Irish and intense than his brothers, with a psychology coiled tightly as a spring, after graduating from the University of Virginia Law School in 1951 Kennedy took a job in the Criminal Division of the Department of Justice. Assigned to menial legwork on tax-fraud cases in a district office in Brooklyn, he quit after only a few months to work on his older brother Jack's senatorial campaign. After Jack was elected in 1952, though, twenty-seven-year-old Bobby felt stranded, a lawyer with no courtroom experience and no certain way to turn. His father suggested that he join Senator Joseph McCarthy's Permanent Subcommittee on Investigations, an offshoot of a low-profile committee on government operations, which McCarthy had turned into a power base for his notorious communist witch-hunt.

Although not yet his own *ism,* McCarthy was already notorious and dangerous. No politician of the age, said the writer Richard Rovere, had "surer, swifter access to the dark places of the American mind."[5] Although he had never been able to prove his charge that 205 Communists had infiltrated the State Department,* McCarthy continued to use

---

ford Carver, who had worked for Onassis during a short-lived whaling venture in California; Nicholas Cokkinis, who ran Onassis's New York office on Broad Street; and Arne Storen, a naval architect and friend). From the point of view of voting stock control, Cokkinis (who had become a U.S. citizen ten days before acquiring his USPC stock), Carver, and Storen held the balance of power; in a dispute between Berenson and the Onassis–Konialidis partnership, any one of the three could put the foreign interests in control. (Senate Report on the Sale of Government-Owned Surplus Tanker Vessels, p. 21, May 29, 1952. See also Memo to J. Edgar Hoover from James M. McInerney, Assistant Attorney General, Criminal Division. April 17, 1952. FBI Report U.S. Petroleum Carriers Inc., May 10, 1952. Fraud Against the Government. Acquired through FOIA.)

* Senator Joseph McCarthy first came to national attention when he was reported in the Wheeling, West Virginia, *Intelligencer* as telling a Republican women's club, on February 9, 1950, "I have here in my hand a list of 205 that were known to the Secretary of

innuendo, and allegations of Communist conspiracies to destroy America, to exploit the fears and frustrations of a nation weary of the war in Korea and worried by Communist advances in eastern Europe and China. Although John Kennedy strongly advised against it, Bobby decided it would be exciting to expose corruption and uncover Communist plots; he told his father to go ahead and talk to McCarthy about giving him a job.[6]

Joe Kennedy had been a financial backer of McCarthy for some time,[7] although how much he poured into the Republican senator's coffers is still not known.* Nevertheless, the amount was apparently sufficient to stop McCarthy from going to Massachusetts to stump for John Kennedy's Republican opponent when the future president was running for the Senate in the Eisenhower landslide year of 1952—a move many believe ensured Kennedy's victory.

Joe Kennedy may thus have felt he had McCarthy in his pocket when he asked him to appoint Bobby chief counsel on his Permanent Subcommittee on Investigations. But McCarthy said he had already promised the job to Roy Cohn, a young assistant U.S. attorney in New York with his own well-established reputation for anti-Communist zeal (he had helped to convict Julius and Ethel Rosenberg as atomic spies). All McCarthy could offer Bobby was a job as Cohn's assistant counsel.

It was more than a passing disappointment to Bobby; irritated at being passed over in favor of Cohn, who was eighteen months younger than Bobby, the young lawyer began showing signs of a bitter competitive streak that Onassis would ruthlessly exploit in years to come.[8]

"Bobby felt he was getting nowhere," an old family friend said, recalling Kennedy's frustration at having to settle for the role of Cohn's assistant counsel. "He was angry and got mad at people all the time. A lot of people thought he was an asshole."[9] His new brother-in-law, George

---

State as being members of the Communist Party and are still making and shaping the policy of the State Department."

* Part of McCarthy's downfall in 1954, apart from booze, was his refusal to explain a financial transaction to a Senate committee investigating his affairs.

Skakel, considered him a "chicken-shit little bastard," and couldn't see
what his sister Ethel saw in "the little prick." [10] Rather more elegantly,
Jackie Kennedy would later tell Onassis that Bobby had "a gift for
estrangement." [11]

But even friends who were used to seeing Bobby take his frustrations
out on others were puzzled by his angry reaction to Onassis's presence at
Pamela Churchill's soiree on that spring evening in 1953. "Their contempt for each other was palpable . . . a definite sense of physical challenge was in the air," said British diplomat Sir John Russell. [12]

Bobby Kennedy and Ari Onassis, men who could bully as well as
charm—short, rumpled-looking men whose clothes always appeared to
be off the rack, no matter how much they'd cost. "I never saw either
one of them in a suit that was pressed, or shoes that weren't scuffed," said
one mutual acquaintance. [13] They must have looked strangely out of place
among Churchill's elegant guests at the Plaza Hotel that spring evening.
Bobby was twenty-seven years old; Onassis was fifty-three, but pretending to be six years younger. Sir John thought it very odd that two men
born a quarter of a century and a world apart should vie with each other
"like a couple of Kilkenny cats." [14]

When Sir John Russell called Churchill the following day, she was
clearly aroused by the clash of egos. "My, those two fellows bared their
tribal qualities last night, didn't they though?" She had never seen
Bobby's eyes icier, or bluer, she said. They were just like his father's "after
Joe has done something truly unconscionable."

Sir John, a man deeply versed in the more scandalous gossip of London society, knew that Pamela's familiarity with the color and chill of Joe
Kennedy's eyes when he had behaved badly was more than mere guess-
work. Later, in Truman Capote's *roman à clef*, Lady Ina Coolbirth would
describe how, when she was Kathleen Kennedy's weekend houseguest, at
age eighteen, she had been raped by Kathleen's father when he slipped
into her bed in the small hours.

The real offense, however, was not Joe's ravishment of her but his failure to leave her some tangible proof of his gratitude. At breakfast, complained Pamela's acquisitive alter ego in the novel: "There was never a

wink or a nod, just the good old daddy of my schoolgirl chum. It was uncanny and rather cruel; after all, he'd had me and I'd even pretended to enjoy it; there should have been some sentimental acknowledgement, a bauble, a cigarette box . . ."[15]

The story was a familiar one to anyone who knew Churchill well. Capote, who was counted among her closest friends, assured Joe Fox, our mutual editor at Random House, that Ina Coolbirth was "not *based on* Pamela Churchill—she *is* Pamela Churchill." *

Is it possible, therefore, that Bobby Kennedy simply lost his cool because he suspected that Onassis—a man who made a point of discovering the indiscretions as well as the weaknesses of his friends—knew about his father's infamous conduct and who knows what other Kennedy secrets Pamela might have let slip? Certainly, despite all his brashness, Kennedy was easily unsettled. But is it a sufficient explanation for their head-on collision at Churchill's party, and what turned out to be the start of a mutual hatred that would fester for the rest of their lives?

"My wife, Aliki, said that she thought each had recognized in the other something he despised in himself, and that rather made sense to me," Sir John would later tell me.[16] But, insightful as Aliki's observation might have been, there was also another and more immediate reason why the two men had taken an instant dislike to each other when they came face to face for the first time.

While Joe McCarthy and Roy Cohn were hunting down subversives in the State Department and making themselves notorious, Bobby Kennedy completed a brief investigation into an alleged influx of homosexuals into the State Department, then embarked on a study of trade—"the blood trade," McCarthy would later call it, with his instinct for a grabby

---

* Capote also told his biographer Gerald Clarke that Pamela had experienced everything that happened to Lady Ina in his story—including being raped by Joe Kennedy (Christopher Ogden, *Life of the Party: The Biography of Pamela Digby Churchill Hayward Harriman,* p. 257. London: Little, Brown & Co., 1994.)

headline[17]—between America's allies and Red China, whose forces were fighting U.S. troops in Korea.

For weeks, his shirtsleeves rolled up, his necktie pulled loose.[18] Bobby had raked through the Lloyds of London shipping index, Maritime Commission records, CIA reports, and Naval Intelligence records.[19] That spring, his revelation that over three hundred ships owned by the New York Greek shipping families were regularly trading with Red China brought him his first national newspaper headlines. It didn't make sense, he declared at his press conference, that "our major allies, whom we're aiding financially, should trade with the communists who are killing GIs."[20]

But patriotism was a meaningless concept to Onassis, and he was furious at what he regarded as the meddling of a politician who was simply in a hurry to make a name for himself.

Although none of Onassis's vessels were involved in "the blood trade," he was concerned that Kennedy's harangues against the Greek shipowners could raise questions about how they had been able to acquire so many of the prohibited T2 tankers engaged in the trade. In turn, this might breathe fresh life into a flagging FBI investigation into his own criminal conspiracy to defraud the United States government with his acquisition of T2 tankers five years earlier.

At any other time he would have lost little sleep over the matter. Even if all his T2s were repossessed by the United States, they had already made him at least thirty times what he had paid for them. Normally, their expropriation would simply have been an inconvenience, but now the consequences could be disastrous, for he was secretly negotiating a deal to supply and run a fleet of tankers for the government of Saudi Arabia to transport its oil under the Saudi flag. It was what he would later call "the biggest dry-throat deal"[21] of his life. The profits were potentially massive, the global implications immense; if he succeeded, he would become richer and more powerful than some nations. Moreover, he also knew that a Saudi tanker fleet would be perceived as the first step toward Saudi self-sufficiency in the oil business. A blatant violation of an agreement between the American oil majors and the Saudi monarchy, it was clear that the deal would have to be concluded swiftly before news of what he was up to got back to Washington and ARAMCO.* Moral principles had played little part in his success so far, but he knew that the Saudis might have second thoughts about their relationship if he were to become embroiled in a Washington scandal.

In October 1953, his worst fears came true. A federal grand jury handed down a sealed indictment against him; now every time one of his ships pulled into an American port, it was seized and its revenues impounded. Nobody was in any doubt about what the sealed indictments contained, and Onassis blamed Bobby Kennedy for his predicament.

"Nobody gave a shit about who owned those tankers until Kennedy started shooting off his big Irish mouth," he told his long-time partner, Costa Gratsos. He was convinced that Bobby had embarked on a personal vendetta against him.[22]

At first Onassis's friends dismissed this idea as paranoia, but that autumn a ten-year-old ONI (Office of Naval Intelligence) report to the FBI surfaced in New York concerning a tobacco insurance scam Onassis had pulled off a decade earlier.

> The tobacco was sent via Genoa, where it was transshipped. And it appears that ONASSIS hit upon the idea of spraying the bales with salt water during their stay at Genoa, the resultant collections from underwriters for sea damage forming a welcome addition to legitimate trading profits. Insurance agents were involved and, eventually, an employee gave the game away at a time when Nicolas KONIALIDIS, brother-in-law of subject,† was in Genoa, with the result that the latter served a term in jail. The dossier appears to have been sent to Greece from Genoa, but is thought to have been lost owing to a close liaison between ONASSIS and one MICHALAKOPOULOS, a Greek minister at that time.[23]

---

* The consortium dominated by four major American companies: Standard Oil Company of California, Mobil, Exxon, and Texaco.

† Also known as Nicos Konialidis. He married his cousin, Onassis's half-sister Merope, in 1938.

Onassis knew that Bobby had trawled through Naval Intelligence records for information with which to attack the Greek shipowners, and, he claimed, Bobby's fingerprints were all over the ONI report which was soon being passed around at smart New York dinner parties like a *samizdat* story in Moscow. Although there was no proof of Bobby's culpability, John Kennedy had worked in World War II for ONI in Washington, and that was evidence enough for Onassis.

"Ari offered Johnny Meyer a Cadillac if he could prove that Jack Kennedy had given Bobby the story," Yannis Georgakis told me.[24]

Meyer, who had been Howard Hughes's Washington fixer in World War II, discovered that Jack had left Naval Intelligence five months before the ONI report to the FBI was written.[25] "I found out Kennedy had been moved to a less sensitive post when the FBI discovered that he was banging some dame (Inga Arvad) they suspected was a German spy,* Meyer would later tell Brian Wells, a former newspaper executive and Palm Beach neighbor, who became his collaborator on an unfinished autobiography.

The time gap didn't convince Onassis, who continued to insist that Bobby had acquired the ONI report through Jack or Jack's intelligence connections. "I said, 'Jesus, Ari, Howard's paranoid, but you're even crazier than he is,' " Meyer would later recall. "He said, 'Even paranoids have enemies. This Irish fuck wants to bury me.' "

A few days later, Onassis asked Meyer if he would get some dirt on Bobby for him. But Meyer was still with Howard Hughes, and didn't want to get any more involved than he already had. "Ari had a lot of charm, a terrific sense of mischief, which I liked, and I had to keep reminding myself that he also had a reputation for ruthlessness. Let's say

---

* "I'm afraid she's dangerous. She certainly has connections with the Fascists in Europe, Germany especially," Jack Kennedy confided to Palm Beach friend, Henry James. "But as for being a spy, it's hard to believe she's doing that, because she's not only beautiful, but she's warm, she's affectionate, she's wonderful in bed. But you know, godammit . . . I found out that son-of-a-bitch Hoover had put a microphone under the mattress!" (Cited in Nigel Hamilton, *JFK: Reckless Youth*, p. 489. New York: Random House, 1992.)

he had a very Greek attitude towards his enemies. If Bobby was going to be hurt—I didn't want any part of that," Meyer told Wells.

Six months after his abrasive encounter with Bobby at Pamela Churchill's party, Onassis flew to Dusseldorf to consult Dr. Hjalmar Schacht about the Saudi deal. Former president of the Reichsbank, and financial mastermind of Hitler's Third Reich, Schacht now specialized in advising countries in the Muslim world. According to the West German magazine *Der Spiegel,* the old Nazi had become the "medicine man of high finance"[26] who was "worshipped almost mystically" through the Middle East, and Onassis calculated that the potential massive profits were worth the risk of further alienating the American government.

By the end of December, Schacht had drawn up a draft contract known simply as the Jiddah Agreement. This closely guarded secret agreement called for Onassis to supply five hundred thousand tons of tanker shipping toward the establishment of the Saudi Arabian Maritime Company, or SAMCO. Headquartered in Jiddah, but exempted from Saudi taxes, the fleet would fly the national flag, and its officers would be drawn from a Saudi maritime college to be established and funded by Onassis. The company would get priority rights on the shipment of Arabian oil, with an initial guaranteed 10 percent of the country's annual output. Although tankers owned and registered in the name of ARAMCO's parent companies on or before December 31, 1953, would not be affected by the deal, Onassis would be free to take over the ARAMCO ships when they became obsolete. A London shipping broker called the arrangement "a carnival of grab." Within a decade, it would have given Onassis a monopoly on the transport of more than forty-five million tons of Saudi oil a year.

On January 18, 1954, Onassis and his young wife, Tina—who was twenty-four years old and the mother of Alexander, aged six, and Christina, four—arrived in Jiddah aboard the tanker *Tina Onassis.* Onassis would later claim that his plan was to approach the oil industry, explain his position, and proceed in consultation with them.[27] But neither ARAMCO nor Washington had been told a thing when, two days later, finance minister Sheik Al Suleiman signed the Jiddah Agreement on behalf of the Saudi Arabian government.

Even so, it still had to be implemented by royal decree, and until the sovereign's signature was on the contract Onassis knew he had nothing. He suggested, therefore, that the deal be kept under wraps until after the royal assent. The aging finance minister willingly agreed. Now that the *ad valorem* had been settled, he was happy to distance himself from the arrangement and leave the timing of the announcement to Onassis.

But unbeknownst to Onassis, brother-in-law and long-time rival— as well as his wife's lover—Stavros Niarchos, knew exactly what was going on.

According to one Onassis biographer, "It was never clear just when, exactly, Niarchos obtained a copy of the document known as the Jiddah Agreement, or by what means he managed to lay hands on it."[28] Nevertheless, there was never any doubt among their friends that it came directly from Tina. Whether it was an act of vengeful betrayal (Tina knew all about her husband's women) or mindless pillow talk, or a childish impulse to cause mischief in the family, is still unclear. But whatever her reasons, it would affect all their lives more profoundly than she could possibly have ever imagined.

TWO

ENTER STAVROS NIARCHOS

What looks like the truth
Is truth enough for fools.

—PAUL CHEVALIER GAVARNI, 1801–1866

Stavros Niarchos was not a man who held much brief for family loyalties. When his mistress Tina Onassis—his wife Eugenie's younger sister—told him about her husband's plans to create a Saudi Arabian fleet, he saw his opportunity not only to sabotage the deal but also to ingratiate himself with Washington, where he faced the same sealed indictments as Onassis for his own illicit acquisition of U.S. war-surplus tankers.

A small, vain man whose habitual expression was that of someone struggling not to sneer, and whose plummy way of talking sounded more like a speech defect than the patrician tone to which he aspired, Stavros Niarchos lacked Onassis's charisma. But he did have enormous cunning, a prodigious capacity for intrigue—and powerful connections. And a few days after Saudi finance minister Sheik Al Suleiman put his signature to the deal that would eventually give Onassis the rights to ship the bulk of the oil flowing out of the Arabian kingdom, Niarchos turned to one of those connections for help to wreck the deal.

Alfred Conrad Ulmer was a tall, gracious southerner whose quiet voice and *savoir faire* concealed a seasoned American spymaster. One of the handful of OSS officers who had never left active intelligence service after the organization was disbanded following World War II, he had stayed on to wind up affairs in Vienna, and hence was on the spot when the Central Intelligence Agency was created in 1947.[1] Now station chief in Athens, one of the most important CIA appointments in Europe, he enjoyed more power than the ambassador.[2]

Niarchos was already regarded as one of the CIA's best assets in Greece. He had gone out of his way to be helpful to Ulmer, facilitating friendships with important Greeks, including Queen Frederika, the manipulative mother of the future King Constantine. Incidentally, Ulmer's boss— the legendary head of the Central Intelligence Agency, Allen Dulles— embarked on an affair with Frederika after they met aboard Niarchos's yacht.

But Niarchos's tip-off about his maverick brother-in-law's Jiddah Agreement was something else; this was intelligence of the highest order. For it was clear that if Onassis succeeded, he would break ARAMCO's (i.e., America's) grip on the region's massive oil supplies, and a strategic asset would be put in serious jeopardy.

Ulmer knew that Niarchos would expect, and unquestionably deserve, a generous *quid pro quo* when the sealed indictments against him were opened in Washington. Nevertheless, he also knew that the Agency would have to tread carefully. Niarchos was having an affair with Onassis's wife, and Ulmer was wary of letting the Agency, or himself, get drawn into the tangled web of a Greek family vendetta. Even so, it was evident that Onassis was deeply ambitious and, as Niarchos pointed out, if he succeeded in Saudi Arabia, what was to stop him repeating the formula in other regions?

Obviously something had to be done, and quickly. It is still unproven whose idea it was to use Niarchos to front the CIA plot to wreck the Jiddah Agreement—a plot that would swiftly become an international conspiracy involving Richard Nixon, the FBI, and the CIA, and escalate from an economic to a political, then finally, to a military confrontation with Onassis. But the decision to destroy Onassis began, according to a former CIA agent who worked for Ulmer in Athens, as "a personal sop to Niarchos from Allen Dulles."[3]*

"Everything between my father and Ari was personal. If Ari didn't succeed in pissing off my father over a deal, he felt it was hardly worth doing," Constantine Niarchos told me. "They goaded each other endlessly. I don't think the Jiddah deal in itself bothered my father too much, but it was a great coup for Ari and he was determined to stop it, just as Ari would have been if the shoe had been on the other foot."[4]

Under orders from the CIA, Niarchos hired former FBI counterintelligence agent Robert Aime Maheu, who had just started his own private-detective agency in Washington, D.C. The Maheu Company was part of an archipelago of investigative agencies set up in the early 1950s by the CIA to circumvent the constraints against domestic operations imposed by its charter. Backed by a monthly Agency retainer of $500,[5] the going salary of a middle-rank CIA officer, the company handled assignments that were "too dirty or too sensitive"[6] for the Agency to involve itself in directly.

Niarchos dispatched his London lawyer, L.E.P. Taylor, to brief Maheu at the office he shared on 15th Street, N.W., with another former

---

* Ulmer himself proved to be a determined source of information on the tensions between Niarchos and Onassis. In the spring of 1995, I visited Ulmer in Washington. The former CIA bureau chief in Athens was now seventy-nine years old; he had suffered a severe stroke and was undergoing intensive physical and speech therapy in a nursing home in Bethesda. In spite of the severity of his incapacity, the old spy had told his son, Alfred Conrad Ulmer, III, that he wanted to talk to me. "I'm not sure how much you'll be able to understand but he seems keen to see you," his son told me as we drove to the nursing home in Montgomery County. Unable to write, able to converse only in monosyllables, Ulmer was frustrated and angry; it was with a tremendous struggle on his part, and intense concentration and patience on mine, that we managed to exchange a few stories about Onassis and Niarchos. (An old pro, he insisted that part of the interview be conducted on "background" rules, which meant that I could use the material but not identify its source. I have respected that agreement.) I still have a picture in my mind of the old warrior sitting in a wheelchair on the manicured Washington D.C. lawn in the bright May sunshine, swapping stories, pumping me for more, refusing to give up.

FBI man, a forensic accountant named Carmine Bellino. Taylor told
Maheu that Niarchos wanted him to scuttle Onassis's contract with the
Saudis and make sure that no oil was ever shipped under the Jiddah
Agreement. Taylor claimed that his client wanted only to protect the
West's interests in the Near East, and sustain the region's *status quo*.
Maheu knew this was risible, but he kept a straight face. "You didn't have
to be Einstein to know that this was about two guys who loathed each
other's guts, who would cut each other's throats for a dollar," Maheu told
me over lunch in the summer of 1996.[7]

Nevertheless, the threat that the Jiddah Agreement posed to the sta-
bility of Near Eastern politics was clearly real. "In just one year," Maheu
told me in his slow, emphatic voice toughened with the granite of work-
ing-class Maine, "Onassis would have controlled more deadweight tonnage than all the Allied nations controlled during the most severe year of
the Korean war; it would have exceeded the kind of power Saddam Hus-
sein would have had if he had won the Gulf war."[8]

Insisting to this day that he was unaware Niarchos had been sent to
him by his CIA paymasters, Maheu solemnly assured his emissary:
"You've come to the right man."

Meanwhile, in London, Niarchos had also hired Alan Campbell-Johnson
to get the matter raised in Parliament, where he believed his animus for
Onassis would find a political echo. The wartime aide to Lord Louis
Mountbatten, the supreme commander of South East Asia Command,
and also his press attaché when he was India's viceroy in the last days of
the raj, Campbell-Johnson had been in the thick of the realpolitik of
Indian partition and independence. This was child's play by comparison.

"I set the whisperers to work at Westminster, got questions asked
in the House. I knew that it was essentially a fight between a couple of
greedy Greeks wanting to do each other down. They had a primitive
relationship: passion, hate, jealousy, envy, all that sort of thing. Although
I believed that it was important on very sound political grounds to stop
Onassis in his tracks, I knew that at the heart of my assignment was
this awful family vendetta which complicated everything," Campbell-
Johnson told me.

In Washington, meanwhile, Maheu had contacted Richard Nixon.

The move to involve the American vice president, a main player on the National Security Council, was brilliant for two reasons. First, an important Nixon function was to raise campaign funds for the Republican Party, and the Onassis–Saudi deal would have upset the bountiful oil companies even more than it had upset Niarchos. Second, Nixon was close to both the Secretary of State, John Foster Dulles, and his younger brother, Allen Dulles.[9] Through Nixon, therefore, Maheu had thrown the hooks into the highest echelons of the State Department and the CIA—though the Agency, of course, had been involved since the moment Ulmer told Niarchos to hire him.

When King Saud intimated that he would ratify the Agreement in spite of American protests, U.S. ambassador George Wadsworth cabled Washington, "High Saudi circles obviously lack background experience to really appreciate questions of principle involved and to their shrewd trader minds it follows that if the foreigner cries he must be hurt financially and they gain in corresponding measure."[10]

Nevertheless, it was Onassis who was regarded as the villain. "We believe it was a case of a smart Greek selling the SAG* a bill of goods . . . Onassis apparently has some mighty plans to monopolize the tanker industry by playing the same theme to Kuwait, Iran, and Iraq," Allen Dulles briefed his brother, Secretary of State John Foster Dulles.[11]

But as the CIA chief stepped up the dirty tricks campaign against Onassis and his brother increased the diplomatic pressure on the king, Nixon was contemplating a faster way of getting rid of the problem. Onassis was the problem, he told Maheu in a furious undertone when the two men met again to review the situation a few weeks after their first meeting on Capitol Hill. The vice president's voice was so soft, his sentences so elliptical and unfinished, that Maheu was frequently unable to make out what he was saying.[12] He only knew the vice president was talking about Onassis—and that "the subject made Mr. Nixon very unhappy."[13]

But as Maheu was about to leave, Nixon abandoned his caution: "And, just remember," he said, gripping Maheu's hand, his voice sud-

* Saudi Arabian government.

denly sharp and clear, his eyes brimming with heartfelt malice. "If it turns out we have to kill the bastard, just don't do it on American soil."[14]

The Justice Department had meanwhile seized more than a dozen of Onassis's ships and impounded their revenues. More convinced than ever that Bobby Kennedy was at the root of all his troubles, Onassis told the Australian journalist Sam White, an old drinking companion in Paris: "Kennedy turned up the heat when everybody else wanted to forget the whole fucking [T2] business. He hates my guts, and I'd hate his if the little shit had any."[15]

Word of the seizure of his first ship by a U.S. marshal came when Onassis was dining at his avenue Foch apartment in Paris with Costa Gratsos and a group of friends. He stood up, poured himself a very large whisky, went to the window and stared out, his shoulders hunched, swaying a little on the balls of his feet, a familiar stance that reminded Randolph Churchill of "a bantam-weight watching which way his opponent is going to move." Gratsos knew it wasn't nerves that made him restless. Anger, not nerves, energized Onassis. The Saudi deal still had to be ratified by the king, and the last thing Onassis wanted was a public brawl with the United States. The more his agony increased, the more he blamed Bobby Kennedy.

Although there appeared to be nothing to link Kennedy to the tanker seizures—except Onassis's paranoia—*prima facie* evidence of a Kennedy connection did exist. Maheu's partner, Carmine Bellino, had been Joseph Kennedy's personal accountant and private investigator before Bobby recruited him to help trace the ownership of the Greek ships involved in the "blood trade" with Red China; he was also the first man Bobby would hire when he became chief counsel of the Senate Rackets Committee in 1957.* "With ledgers and cancelled checks spread out before him, (Bellino) could reconstruct the most carefully disguised financial transac-

---

* Formally known as the Senate Select Committee on Improper Activities in the Labor or Management Field, it was launched in January 1957 to investigate Teamsters leader Jimmy Hoffa and Labor racketeering.

tions," said Kennedy family friend and Bobby's biographer, Arthur Schlesinger.[16]

Although he still had no proof, Onassis continued to insist that Bobby was working to thwart him. He went back to Johnny Meyer. A man who had explored the limits of Washington graft and had even been a principal witness in a Senate investigation into government corruption, Meyer knew how things were done there.

"Ari said, 'Johnny, I know Bobby Kennedy wants to destroy me. I just want you to prove it,' " Meyer later told Brian Wells. This time, perhaps because he suspected that there was some validity to Onassis's obsession with Kennedy's hostile intent toward him, Meyer agreed to give Bobby "a bit of a spin."[17]

A few weeks after Meyer began his investigations, the CIA ordered Maheu to end his partnership with the once and future Kennedy cohort Carmine Bellino.

"Because of Bobby, the CIA told me that if I were to work with the Agency, I would have to move away from Carmine and any possible Kennedy connection," Maheu later explained as the reason why he had been forced to dump his partner shortly after getting the lucrative assignment to sabotage Onassis's Saudi deal.[18]

He knew that there had been a deep animosity between the Kennedys and the CIA since the time Bobby had worked for Joe McCarthy, who, in Maheu's own words, had gone after the State Department like a "dog after a bone. State was run by John Foster Dulles. The CIA was run by his brother Allen. Like much in politics, the CIA's ideological stand was built upon a very personal foundation."[19]

This might have been true as far as it went, and Robert Maheu might even still believe it, but the Englishman Campbell-Johnson suspected that there was a more compelling reason why Maheu had been ordered to cut Carmine Bellino adrift after Meyer began sniffing around in Washington.

"The conspiracy against Onassis was a very odd brew of interests, and perhaps the truth will never be completely untangled," Campbell-Johnson told me in London late one February afternoon in 1997.[20] "My own view is that the CIA made Maheu get rid of Bellino not to remove a link to Bobby Kennedy but to *establish* one—to convince Onassis, when Johnny Meyer discovered the connection, as he almost certainly would do, that Bobby *was* running the show." *

I understood how Onassis's devious mind would jump to that conclusion. But why did the CIA want to implicate Bobby Kennedy? Why did they take such elaborate pains to persuade Onassis that Kennedy was, in Campbell-Johnson's imperial rhetoric, "the dusky fellow in the woodpile"? Why would they go to such lengths?

Campbell-Johnson had been at the heart of the conspiracy to destroy Onassis's Jiddah Agreement, if not Onassis himself; I am not in the least certain that he did not also have ties to British intelligence; and his views on the CIA's purpose cannot easily be dismissed.

Imagine, he told me that February afternoon over lunch, how it must have felt to be in Allen Dulles's position at that moment? If Onassis's notion about Bobby Kennedy were not swiftly confirmed, his suspicion might next have fallen upon his brother-in-law, Stavros Niarchos.

"I don't imagine that Dulles cared a jot whether Onassis found out about Stavros's part in the plot to smash his Jiddah Agreement; nor was it a question of protecting a CIA asset, which Stavros undoubtedly was," Campbell-Johnson said.

Dulle's concern, he suggested, was for his own safety. For the CIA chief had embarked upon an affair with Queen Frederika of Greece, which Niarchos had set in motion when he introduced them aboard his yacht. What use Onassis could have made of that piece of intelligence! Understandably, then, the last thing Dulles wanted was for Onassis to start nosing around any more than usual in Niarchos's affairs.

Bobby was the perfect decoy. And the irony was that the CIA hated him and the Kennedys as much as Onassis did. "Of course, I can't prove it was Dulles himself who threw Bobby to Onassis, although I've always believed that that is the real measure of its truth," Campbell-Johnson said.

---

* The conspiracy to destroy the Jiddah Agreement.

*   *   *

Meanwhile, Costa Gratsos had persuaded Onassis that Bobby could wait. First, he should confront the attorney general, Onassis's former lawyer Herbert Brownell, whose advice six years earlier had helped him circumvent the law and acquire the war-surplus tankers. He argued that Brownell could not be feeling too comfortable at the prospect of being interrogated on the stand about his attitude toward the law before he became attorney general. "Let's find out exactly what's in those sealed indictments, because whatever Brownell knows about you, you certainly know plenty about Brownell," he said. If he returned to the United States now, Onassis would have an edge: "The Justice Department won't be expecting you, and Brownell won't want you," Gratsos told him.[21]

On February 1, 1954, Onassis returned to America and waited to be arrested. But by February 4, he was still free and his patience was exhausted. If he was under indictment, why hadn't he been arrested? He sent Brownell a telegram:

> Honorable Herbert A Brownell Jr. Attorney General of US Department of Justice Washington I wish to inform you that having arrived from Europe on Monday night I place myself at your disposal during my visit to this country for any information you or your department might care to have—Onassis.[22]

The following day, Friday, February 5, he was arrested while lunching at the Colony in New York, and charged with criminal conspiracy to defraud the United States government; fingerprinted, mug shots taken, he was briefly put in a holding pen with muggers, male prostitutes, and a posse of Puerto Ricans charged with shooting up Congress.

Onassis's troubles continued to mount.

A letter J. Edgar Hoover had written twelve years earlier surfaced in the *New York Daily News* claiming that Onassis had "expressed sentiments inimical to the United States war effort, and that his activities and movements while in the United States should be carefully scrutinized."

Spyridon Catapodis, Onassis's middleman in the Jiddah deal, filed suit in criminal court charging that "by various fraudulent maneuvers"—including signing documents with invisible ink!—Onassis had swindled him out of $200,000. In a sixteen-page affidavit, sworn at the British consulate in Nice, he also disclosed the details of the graft Onassis had paid to palace aides, including $350,000 to Finance Minister al Suleiman.[23]

But bribes are a fact of life in desert kingdoms, and it was the tone rather than the details of the affidavit that worried Onassis, who was portrayed as a man who would not hesitate to double-cross the Saudis if the Americans came up with "either a large amount of money or some other valuable concessions" to make it worth his while to break the contract.[24]

Perhaps Catapodis was scared of the political repercussions, as he claimed; perhaps he did urge Onassis not to mess with ARAMCO, as he also claimed; and perhaps Onassis did tell him that he intended to become "the richest and most powerful man in the world" no matter who got hurt in the process. And perhaps Robert Maheu believed every word of it when he anonymously sent copies of Catapodis's allegations to oil men, politicians, and rival shipowners, "to encourage the rumors and keep Onassis guessing where the next blow was coming from."[25]

Soon it was clear that Onassis was fighting for his very survival. More and more of his tankers were laid up as oil companies canceled regular contracts; accused of violating Peru's territorial waters, nine of his whaling ships were commandeered by Peruvian marines, and his factory ship was raked by machine-gun fire.

Even though Foster Dulles knew that ARAMCO's decision to boycott the proposed Onassis–Saudi tanker fleet could result in the Saudis nationalizing the whole industry, he told King Saud that the United States would back the oil companies to the hilt in any showdown. Dulles advised the monarch that he should bear in mind what had happened to prime minister Mossadeq—toppled in a CIA-assisted coup in favor of the young Shah Mohammad Reza Pahlavi—when he had played the nationalization card in Iran.[26]

In October, the king backed down. Onassis's dream of creating a Saudi–Onassis superfleet was over. The end of Jiddah confirmed what

Onassis had always believed: In America, everything is connected—and he was more convinced than ever of how powerful an enemy he had made in Bobby Kennedy.

On New Year's Eve, Onassis gave a dinner party at the Hotel de Paris. It had been an exhilarating year, he told his guests. "My mistake," he said of the Jiddah misadventure, "was that I woke up too early and disturbed those who were still asleep, and as a result I got into the biggest mess of my life." [27]

But he had taken on the U.S. State Department, the CIA, Richard Nixon, J. Edgar Hoover and the FBI, the American oil giants, Stavros Niarchos, the Dulles brothers, and, he was more convinced than ever, Bobby Kennedy—and he had almost beaten the lot of them, he added triumphantly. It was time to move on. While his mind had been on the Saudi deal, his plans for Monte Carlo had barely been mentioned. Now he was determined to turn that "sorry-ass town into something very tasty." [28]

Few realized that the boast hid the fact that he was in bad shape financially (Niarchos was telling friends that Onassis was down to his last $5 million), and that Ari faced the bleak prospect of celebrating his next birthday in bankruptcy. [29]

THREE

"GIVE ME AN HONEST WHORE ANYTIME"

Except for the prince of Monaco, whom the French regard as an inconsequential person whose only real interest is in a reliable source of funds for his pleasures, Onassis may now be regarded as the real ruler of Monaco.

—J. EDGAR HOOVER[1]

I t was not an easy summer for Onassis, and Monaco still offered him his best chance of rebuilding his fortune. Yet if Prince Rainier were to die without a direct male heir to succeed him, under the Treaty of Protection the principality would come directly under French rule—and along with it, the unwelcome exigencies of French taxes. With a healthy young man of thirty-three on the throne, this should not have been a problem for his people.

However, a recent bank scandal, which had threatened the principality with bankruptcy,* had focused their attention on how vulnerable the state was to the vagaries of fate, as well as the judgment of princes.

Onassis worried about this a great deal, but found it almost impossible to discuss rationally with the prince. The friction between them was complicated not only by their huge egos but also by what seemed to be an extraordinary reversal of their expected roles: The prince wanted to turn Monte Carlo into a Las Vegas–style operation; Onassis wanted to restore its lost glory in order to create a sanctuary for the rich.

Onassis had little time for the casino side of his investment; in fact, he considered "that kind of gambling" immoral, he had told Rainier shortly after taking over the SBM. "I don't need you to tell me what is moral or immoral, Mr. Onassis," answered Rainier loftily, to which Onassis snapped, "If you're going to be a woman of ill repute and make no money, you might as well be an honest woman."

But the two men resented the way the press gleefully fanned the flames of even their most petty quarrels, and both tried to leave each other alone as much as they could. And so it was to Rainier's American chaplain that Onassis turned to discuss the hereditary crisis.

The Very Reverend Francis Tucker, a peppery sixty-three-year-old who relaxed during afternoons on the beach in a clerical-black bathing suit, and mixed martinis with as much aplomb as he performed matins, did little to disguise his dislike of Onassis, whom he regarded as a "claim-jumper, anxious to plant the Panamanian flag atop the palace."[2]

---

* Two years earlier, finance minister Arthur Crovetto, the head of one of the oldest families in Monaco and the director of Rainier's cabinet, broke the golden rule that the principality's reserves be spread among the state's thirteen banks and had deposited the lot in the Societe Monegasque de Banque et de Metaux Precieux. The bank had speculated massively in a television company that had run into trouble. Crovetto invested a further 900 million Treasury francs to boost the stock, but in three days on the Paris Bourse its value crashed more than fifty percent. Crovetto pleaded with the National Council—eighteen businessmen elected by the principality's six hundred male voters—for 330 million francs to avert the collapse of the state. Having been ignored by the palace clique for so long, the council members agreed to cover the debt on the condition that Crovetto was sacked, along with the rest of Rainier's cabinet cronies.

The two men met aboard Onassis's yacht. Clutching a glass of scotch, Tucker listened to Onassis's concerns. His mood had become mellow when Onassis suggested that Rainier's "symbolic value would accrue wonderfully" if he had a beautiful princess with whom he could share the heavy burdens of princeship.[3]

Onassis knew that Tucker had more than pastoral clout at court: At least one royal mistress had slipped out of the picture after he had greased the path. "I suggested he look in his own country: An American princess would be romantic and visionary—the way Americans like their dreams. I thought perhaps Marilyn Monroe," Onassis would recall.[4]

Even a man far less experienced in the complexities of a life at court than the Vatican-trained priest would have given a wide berth to this idea. But religion, money, and politics lead to strange alliances, and Onassis was certain that Tucker had given him a tacit nod to explore the possibility with the star whose marriage to the legendary American baseball hero Joe DiMaggio had just ended.

Onassis called Spyros Skouras, whom he had known since the forties, when Skouras served as the chairman of the Greek War Relief Fund in California. Also Monroe's Hollywood patron and occasional lover, he was now the head of her studio, 20th Century Fox. But Skouras advised Onassis was to forget the idea. Monroe was a mess, he said. She was living with the playwright Arthur Miller; she was having an affair with a rich young senator named Jack Kennedy, who was jumping from one Hollywood starlet's bed to another. "Mahlin's a won'erful girl, a switthar, but she ain't no princess," Onassis recalled him saying. Besides, Monaco was smaller than his backlot at Fox. "Mahlin'll want a bigger stage than that, fahcrissakes."[5]

Onassis's idea of casting seems risible in retrospect, and to many it seemed risible then. Nevertheless, he called Greta Garbo's lover Georges Schlee, a Russian social fixer, who brokered an appointment with Monroe through Gardner ("Mike") Cowles, Jr., the publisher of *Look* magazine and Monroe's neighbor in Connecticut.

The role appealed to Monroe immensely, although she was somewhat vague about Rainier's identity. "Is he rich? Is he handsome?" she asked.

Cowles didn't think she even knew where Monaco was. Did she think, he asked cautiously, the prince would want to marry her? "Give me two days alone with him," she answered with the insouciance of a star who knows that her name will always appear above the title, "and of course he'll want to marry me."[6]

Onassis's joy was fleeting, however. Schlee had barely given him the good news when Rainier sprang his own surprise: He planned to marry Grace Kelly. The couple had met in 1954, when Kelly was attending the Cannes Film Festival. The magazine *Paris Match* arranged a photo opportunity for her to visit the palace and be shown around the royal gardens and private zoo by the prince. Kelly was a far safer bet for the princess part; the complete antithesis of Monroe, she looked as if she had stepped straight out of the Philadelphia Social Register, although "the stupendous upward mobility of the family"[7] had not carried her quite that far. Her father, John B. ("Jack") Kelly—the son of a County Mayo farmhand who had made a fortune in U.S. real estate and begat a large and boisterous family—seemed uncannily like Joe Kennedy to Onassis, who was beginning to feel bedeviled by "upstart fucking Irishmen and their brood."[8]

Nevertheless, it was with proprietorial satisfaction that Onassis watched the U.S. liner *Constitution* approach Monaco one April morning in 1956 as the band of the royal household played "Love and Marriage" on the quayside, and his seaplane showered the harbor with a thousand red and white carnations. All of this seemed to make Onassis's life a little more promising, for although the marriage was as real as a fairy tale, it was a fairy tale that towns like Monte Carlo thrive on.

But at almost the same time, another marriage, one which would affect the course of Onassis's life even more significantly, was coming unraveled in America.

Jacqueline Lee Bouvier knew all about Jack Kennedy's reputation and had always known what to expect from a man twelve years her senior, who had taken no interest in marriage until his single state began to impede upon his political ambitions. As she had remarked in her pensive child-woman voice on the eve of their wedding three years earlier: "I just don't think there are many men who are faithful to their wives. Men are such a combination of good and evil." [9]

But if her acceptance of the kind of married life she was letting herself in for seemed somewhat casual, there had been a reassuring *quid pro quo:* Adhering scrupulously to her mother's advice, she was marrying for money.

This is not to say that she did not love her husband, for she did; or that she did not have a powerful sense of herself, because she had that, too. But determined to live a full and glamorous life, she had found a man her own father himself might have picked for her.

And John V. Bouvier III, a rakish New York stockbroker, was the key figure in Jackie's life. "Black Jack," as his Wall Street associates called him, was known as much for his swarthy complexion as for his dark powers of seduction.

Black Jack had never remarried after Jackie's mother Janet Lee divorced him. It had been a difficult marriage from the start, short on love and understanding, and eventually money too: He turned a $750,000 inheritance into a $100,000 estate at the end of a lifetime of wheeling and dealing. As a single parent, he had dominated and indulged Jackie and her younger sister, Lee, well beyond his shrinking means. The result was that although both girls grew up with their mother's firm belief in marrying for money (Janet herself went on to marry Standard Oil millionaire Hugh Dudley Auchincloss, Jr.), they also continued to be attracted to men with something of Black Jack's chauvinism in their makeup: older men, men with style, seductive charm, dangerous flaws in their character—and no pretense of faithfulness at all.

And so, having acquired everything she thought she needed out of a marriage—money, position, and the glittering carapace of social glamour—Jackie had *wanted* to be kept in the dark. Knowledge had been her biggest threat; the less she knew, the more smoothly she believed the marriage would run. But by 1956, her willful blindness to Jack's unfaithfulness was harder to sustain. Those who knew her well sensed that behind her inscrutability she chafed at being required to behave as if everything was fine with their marriage. Truman Capote, a regular guest at their New York apartment in the early days of their marriage, said: "She was sweet, eager, intelligent, not quite sure of herself, and hurt—hurt because she knew that Jack was banging all those other broads." [10]

She already had divorce in mind when she accompanied Jack to California that January, where he was to discuss the narration for a short film on the history of the Democratic Party. The film was to be screened at the opening of the national convention in Chicago in August. [11] Charles Feldman gave a small party for them at his home.

One of the most powerful agents and producers in Hollywood, Feldman had been introducing Jack to actresses for at least a couple of years and knew that the rising young Massachusetts senator's marriage had not dimmed his appreciation for Hollywood pleasures. [12] But on this occasion it was the senator's pretty young wife whom Feldman would introduce to a movie star.

"Bill's a Republican, but don't hold that against him," he told her, taking her over to meet William Holden. "Why should I?" he later recalled her replying. "I have been a Republican all my life." [13]

Holden was one of Hollywood's biggest stars. Married to the same woman for fifteen years, father of two young boys, involved in good causes and civic activities, he was also one of the movie capital's most respectable citizens. But there was a darker side to his public image. Secretly a heavy drinker and a discreet womanizer, at thirty-seven, his brow already seamed with ridges of tension, he might well have reminded Jackie of her father.

"The thing about adultery," he would later tell a friend, "is that you must never get involved with a woman who has less to lose than you have." Playing by such strict rules had likely helped keep him above suspicion and out of trouble for so long.

Jackie, the irreproachable wife of a rich young senator marked out for the highest office, and whose discretion was as absolute as his own, fit the bill perfectly. And at Charlie Feldman's house on Coldwater Canyon in Beverly Hills, they embarked on an affair that according to Gore Vidal (a tenuous relation to Jackie by marriage) was essentially and primarily "motivated by revenge on Jack." [14]

Jackie must have relished the sublime sense of symmetry in this, her first extramarital affair: She was sleeping with Holden in the same bed in which her husband (and others) had made love to Marilyn Monroe. [15]

It was an affair waiting to happen. Few men were as wrapped up in masculine values and macho displays as Jack Kennedy. For three years Jackie had stood by knowing that he was, in Truman Capote's words, "banging all those other broads." How long could she go on looking the other way? Stories of Jack's fondness for group sex, how he loved to watch two women get it on, and his lackluster performance in one-on-one encounters were humiliatingly familiar among their friends. [16]

Jackie knew that he had even talked to women about what she was like, and what she liked, in bed, and how he liked to encourage the exhibitionist streak in her. (Capote told friends how she would invite him into her bedroom while she was dressing to go out with Jack for the evening. When Capote's indiscretion was repeated to her, it ended their long friendship. [17]) Betty Spalding, the wife of one of Kennedy's closest friends, expressed dismay at the bedroom intimacies Jack confided to her. [18]

Was it with these thoughts in mind that a few days after they left Hollywood Jackie told her husband about her fling with Bill Holden? This is what Onassis would later tell friends, and subsequent events appear to confirm it.

Although it would be easy to ascribe her confession to the prompting of her Catholic conscience, it is also possible that her motive was more calculating than a conflict of religious principles. For she knew that Jack got off on women's sexual adventures. Women who had slept with powerful men had always turned him on. He could not be in the same room with a woman with a past without getting an erection, Onassis claims that Jackie later told him. And certainly it was what would attract Kennedy to Judith Campbell (later Mrs. Exner); she had been Frank Sinatra's lover and was the mistress of Chicago mobster Momo Salvatore (Sam) Giancana.

But if Jackie's decision to put her cards on the table about her affair with Holden was a desperate bid to revive her husband's interest in her—to remake herself into the kind of woman she knew he found irresistible—it was both understandable and tragically reckless. Although whether her adultery had the desired effect on her husband's libido can only be guessed at, it was almost certainly short-lived. For not long after returning from California, she discovered that she was pregnant.

As uncertain of whose child she was carrying as Kennedy was convinced that it was not his, only two choices lay before her, both equally disturbing to contemplate.

In August, even though she was nearly eight months pregnant, and despite a particularly grueling Chicago heatwave, Jackie agreed to accompany her husband to the Democratic Party convention. But despite this show of loyalty to him—whether it came out of a sense of guilt or was meant to be her valedictory performance as Mrs. John F. Kennedy—Jack remained unforgiving about the child she was carrying.

Immediately after the convention, disappointed at his failure to win the second spot on the Adlai Stevenson ticket,* he flew back to New York with Jackie and left her to stay with her mother and stepfather at Hammersmith Farm, the Auchincloss estate in Newport. Meanwhile, Jack went off to join George Smathers, his closest social friend on Capitol Hill, for a yachting trip on the Mediterranean.

No one could have failed to see that there was something seriously wrong between the Kennedys. But even if the marriage was on the rocks, Jack's decision to go to Europe when his wife, who had already suffered one miscarriage, was about to have their first child, appalled his friends, including Smathers. "Now that she needed him he seemed to be deserting her," Smathers would later admit.[19]

According to one report, the cruise quickly turned into a bacchanal; another called the yacht a floating bordello. Among the women they brought on board at Nice was "a stunning but not particularly intelligent

---

* Senator Estes Kefauver of Tennessee was chosen.

blonde who didn't seem to have a name but referred to herself in the third person as Pooh."[20] Still, so entranced was Jack with Pooh that he decided she was too good to leave even when word reached the yacht that Jackie had undergone an emergency cesarean operation and given birth to a stillborn daughter.

Of course, it was more complicated than that. Although Jack's anger at Jackie's unfaithfulness was overwhelming, his male pride was even stronger. Unwilling to admit to anyone that he suspected the child might not have been his, he told Smathers: "If I go back there, what the hell am I going to do? I'm just going to sit there and wring my hands."[21]

All his friends knew that where sex was concerned he was capable of the most amazing selfishness. But such brutal indifference to his young wife's feelings at the traumatic loss of a nearly full-term baby astonished even those who had long ago stopped being surprised by his intemperance. Politically, too, it was an inexplicable lapse of judgment. It took Smathers three days to convince Jack to return to the States: "If you want to run for President, you'd better get your ass back to your wife's bedside, or else every wife in the country will be against you," he said.[22]

Kennedy flew back to America the following day.

But when Jackie left the hospital she went to the Auchincloss estate, not their own home. Crushed by the loss, she couldn't, she told friends, bear the sight of the nursery she had prepared for the new baby at Hickory Hill, their Georgian mansion on the Virginia side of the Potomac River. "They were both bitter," recalled Lem Billings, a close friend of the Kennedys, "disillusioned, withdrawn, silent as if afraid that conversation would deepen the wound."[23]

Only the two families knew just how close to divorce the couple were. Jackie's cousin Edie Bouvier Beale later claimed, "My poor cousin, who wanted to divorce [Jack], was not able to, because Mr. [Joe] Kennedy said they had to get into the White House. The whole thing was perfectly awful."[24]

In November Jackie flew to London to stay with her sister Lee, whose marriage to Michael Canfield was in an even more unhappy and precarious state than her own. The handsome adopted son of publisher Cass

Canfield and his social register wife Katsy, Michael was rumored to be the illegitimate son of the Duke of Kent, the younger brother of King George V and Queen Mary—no small cachet in a society where royal bastardy was still considered *chic.*

It was no secret that Lee relished the entrée her husband's position— a minor diplomatic post with the American embassy—and putative royal connection gave her to London's social elite. No longer even bothering to hide her indiscretions, she was conducting simultaneous affairs with a number of its most distinguished members: David Somerset, the heir of the Duke of Beaufort; Robin Douglas-Home, nephew of an archetypal Tory prime minister, lover of Princess Margaret, and all-around black sheep of the family; Lord Lampton, the husband of one of Lee's best friends; and Stanislas Radziwill, a Polish aristocrat who steadfastly stuck to his defunct title of Prince.

What Jackie made of this is unclear. But it is hard to imagine that even in her own unhappy abstraction she could have been totally blinded to the extramarital ring dance going on around her. She certainly did not hesitate to throw herself into the whirl of smart London parties and country house weekends as if she were already a free woman.

The shockwaves her exposure to Lee's fast set made on the Kennedys back in Boston cannot be overstated. Joe Kennedy knew all about the nocturnal intrigues that went on at English country house weekends: He had engaged in plenty of them himself when he was American ambassador to the Court of Saint James. Pamela Churchill, for one, had asserted his involvement most roundly.

When Drew Pearson drew attention in his *Washington Post* column to the ball Jackie was having without her husband in London, Joe—fearful that a divorce would finish his son's presidential hopes—stepped in "to broker a treaty to save the marriage." [25]

Exactly how much this treaty cost is not clear—one million dollars? as *Time* speculated it took; one biographer claimed it required a trust fund for her children, to revert to Jackie after ten years if she remained childless. [26] Whatever the carrot, it was evidently sufficient to prevent Jackie's walking out.

*   *   *

None of this escaped the attention of Onassis. Since the destruction of his Jiddah dream, with its promise of untold wealth, he had been tracking the Kennedy fortunes, biding his time for an opportunity to take his revenge. For although he pretended to be philosophical and self-deprecating about his failures and misfortunes, he never forgave those who hurt him. He rarely instituted lawsuits; he would never give that satisfaction to an enemy, a rival, or anyone else who had gotten the better of him in some way. But thoughts of revenge were never far from his mind.

By this time, Onassis had discovered Stavros Niarchos's part in the conspiracy to break his Saudi deal, despite Allen Dulles's efforts to convince him that Bobby Kennedy was the chief architect of its collapse. "Onassis had never been sure which of the two it was whom he hated most—Bobby or Stavros," Yannis Georgakis would later tell me. Nevertheless, the knowledge of their apparent complicity in his unhappiness made Onassis angry in an exhilarated way.

A patient plotter, he was sustained by the troubles of his enemies. From her London spree, which had cost Joe Kennedy so much, he knew that Jackie Kennedy was trouble, indeed.

Onassis liked to think he was a connoisseur of women, and what he called "classy tarts" were always more trouble than they were worth, he told Georgakis with an uncanny intimation of what was to come. "Give me an honest whore anytime," he said.

Meanwhile, Joe Kennedy's money seemed to have saved his son's marriage—and kept Jack on course for the White House—but the damage had been done. And although the couple pulled out all the stops to conduct themselves in public as if nothing had happened, in reality nothing would ever be the same between them again.

FOUR

A CARNAL SOUL

Big mistakes usually start real small.

—ARISTOTLE SOCRATES ONASSIS

Deals defined Onassis, business gave expression to his very soul, and the collapse of the Jiddah Agreement cost him far more than money.

Yet Tina, a creature of frivolous impulse and promiscuous pleasure, appears to have been let off remarkably lightly by Onassis, even though her betrayal—deliberate or otherwise—in telling her lover Stavros Niarchos about the promised Saudi cornucopia had started the disastrous train of events that led to its downfall. Indeed, Onassis later told Yannis Georgakis that discovering his wife's treachery had sexually aroused him. Georgakis believed that this was the key to their relationship.

Nevertheless, it is also a fact that Onassis still found it easier to blame Bobby Kennedy for his troubles. He believed—or wanted to believe—that the Saudi deal collapsed not because Tina and Stavros had made a cuckold of him but because Kennedy wanted to make a name for himself.

Even so, Georgakis remained skeptical about a connection between Bobby Kennedy's "blood trade" investigation for Joe McCarthy and the eventual failure of the Jiddah Agreement. But revenge had always been a particular pleasure for Onassis, and the anticipation of it was so satisfying that nothing Georgakis said could convince him that he might have gotten Bobby Kennedy wrong.

Meanwhile, although Spyridon Catapodis's fraud case against Onassis had collapsed at the pretrial hearing in New York.* Onassis continued to count the cost of Jiddah: dozens of lucrative U.S. charters were canceled, and more than half of his heavily mortgaged fleet remained idle. "You're looking at a desperate fellow," he told an English friend in the summer of 1956." [1] In Washington, he was fined $7 million for his illegal acquisition of the war-surplus tankers, but in return for guilty pleas, the criminal charges against him were dropped.

Only his burgeoning friendship with the aging British statesman Sir Winston Churchill seemed to lift his spirits—partly because he knew that the friendship annoyed Rainier. The prince was sore at the Brits for having sent only a middling-rank diplomat to represent the House of Windsor at his marriage to Grace Kelly, a snub that had inclined the other royal houses of Europe to a similarly cool response.

In Washington, meanwhile, the U.S. government had become concerned enough about Egypt's president Gamal Abdel Nasser's growing closeness to the Soviets to withdraw its offer to finance the building of a massive dam across the Nile at Aswan. Six days later, Nasser nationalized the Anglo-French owned Suez Canal.

Randolph Churchill tipped off Onassis that British prime minister Anthony Eden was planning to retake the Canal by force in a joint military operation with the French and Israelis. This was priceless information for Onassis, and so secret that even Eisenhower, preoccupied with his second-term presidential election campaign, was not informed by his European allies.

---

* When Onassis's lawyers began to probe the Niarchos-Maheu-Catapodis connection, Niarchos claimed that serious security matters were at stake, and that he "would have to be advised by Washington whether I'm going to answer these questions or not." The hearing was adjourned; four days later, Catapodis dropped the action.

If war did break out in the Middle East, oil supplies to Europe and the United States would be forced to go around the Cape of Good Hope, a journey twice as long as the Canal route. But could Onassis trust Randolph Churchill? It was no secret that Randolph despised Eden and would not hesitate to start any rumor to discredit his father's successor. "Randolph's a drunk but he's the best informed drunk in London," Onassis said when Costa Gratsos raised the question of Randolph's reliability.[2]

In September, Onassis began to redeploy his idle tankers. And when the Franco-British-Israeli invasion came at the end of October, he was the only major shipowner with the best part of his fleet available and strategically placed to cash in on the crisis. It was, Sir John Russell later told me, "a beautiful moment for Ari," as he contemplated the rush of offers from the same U.S. oil giants who, a few months earlier, had conspired to squeeze him out of existence.

Stripped of American support, the Suez invasion collapsed in a humiliating climb-down by the European allies—but not before Nasser had scuttled hundreds of ships in the canal, effectively blocking the main route for western Europe's oil for months to come. Beneath the headline *THE VICTOR OF SUEZ*, the German magazine *Revue* published a full-page picture of Onassis. He sent the picture to Randolph Churchill, inscribed: *"For my favourite spy!"* Attached was a Swiss banker's draft for what one Onassis aide called "a very handsome sum." It was typical of Onassis's ambivalence toward his hero's son that even when he was rewarding him so generously, he could not resist reminding him that it was for betraying his country.

Still in vigorous middle age, and again a millionaire many times over, Onassis regained his old buccaneering spirit. When Greek prime minister Constantine Karamanlis invited him to take over the national airline, TAE, for a mere two million dollars, he leapt at the chance, albeit only after squeezing breathtaking concessions out of the government. In return for his promise to turn the moribund airline—which he had shrewdly renamed Olympic Airways—into a world-class carrier, the government guaranteed to cover his losses on transatlantic flights, compensate him for losses incurred by unofficial strikes, and exempt him from corporate taxes and landing fees in Greece. He was also given a call on government loans of up to $3.5 million at a fixed interest rate of 2.5 percent, and the right to export his profits. One German newspaper summed up his triumph with the headline: HERR ONASSIS KAUFT DEN GRIECHISCHEN HIMMEL (Mr. Onassis Buys the Greek Sky). If he had not quite managed to do that, he was nonetheless the only private citizen in the world who owned a national airline.

Despite this unique achievement, and despite his friendship with Sir Winston Churchill—which he exploited to the hilt—Onassis had lost none of his shady charisma. To Rainier's chagrin, the *Christina,* which dominated the Monte Carlo harbor, was still reckoned to be where the real power in Monaco resided, and remained the mecca for the rich and famous.

When Jack and Jackie Kennedy arrived in town the following summer, Onassis asked them aboard for drinks. The invitation apparently had been Churchill's idea; he had known the family since Joe Kennedy was the American ambassador in London. "They tell me this young man is presidential timber. I should like to meet this presidential timber," Onassis would recall of the old warrior's request.[3]

Unfortunately, as it happened that day, Churchill was having one of his more confused spells and had no idea who Jack Kennedy was. (Kennedy, who revered the old statesman, was disappointed and not a little hurt when the great Englishman disdained his attempts at conversation. "Maybe he thought you were a waiter, Jack," Jackie suggested helpfully afterwards.[4]) Onassis was delighted at the opportunity to snub the Kennedys. "I must ask you to leave by seven-thirty," Onassis dismissed the couple with patronizing regret. "Sir Winston dines sharp at eight-fifteen."[5]

Jackie, however, impressed Onassis very much. Years later he claimed to recall everything about her that evening, from the clothes she wore— "she had on a white, very simple, very expensive suit"[6]—to the way she held herself. "She had a withdrawn sort of quality," he told Hélène Gaillet, who would become one of his closest female friends in the last months

of his life. "It wasn't shyness, it wasn't boredom either. She wasn't conspicuously friendly, but she had a way of making you look at her," he said.

Perhaps, suggested Gaillet, remembering the shaky state of Jackie's
marriage at that time, it was unhappiness that she exuded.

"Sometimes unhappiness in a woman can be sexy," Onassis replied.[7]

The fact is, he would have known a great deal about Jackie before she
even stepped aboard the *Christina* that evening. He regularly demanded
files on his guests; he even kept files on the whores he regularly had
flown in from Paris and New York. But he would not have needed a file
to remind him of the incident that had almost ended her marriage, and
Jack's hopes of the U.S. presidency, the previous year.

Her husband's chronic womanizing was hardly a well-kept secret,
even before he was seen running around the Mediterranean "on the
chase" while Jackie was at home expecting their first child. Rumors of
how much it had cost her father-in-law to dissuade her from seeking a
divorce had been as rife on the Riviera as they were in Washington.

The fact that Jackie could be bought would also have excited Onassis.
Women he could buy were always a reliable aphrodisiac. "There's something damned *willful* about her, there's something provocative about that
lady," he had told Costa Gratsos after the *Christina* reception. "She's got
a carnal soul."[8] Exactly how he had divined this is unclear since another
guest that evening remembers only Jackie's "pleasant diffidence . . . her
little voice" repeating the mindless small talk people fall back on at such
moments.

Nevertheless, Gratsos sensed the complexity and depth of Onassis's
interest in the senator's wife. He guessed what was on Ari's mind—
seducing Mrs. Jacqueline Kennedy, what sweet revenge on Bobby and
the family that would be—and promptly told Ari to forget it. "Don't
fuck up her life just to get even with Bobby," he would later recall telling
Onassis. "Anyway, she's too young for you."[9]

Jackie was just four months younger than Tina.

In 1959, a series of palace maneuvers unseated Onassis's representative
from the helm of SBM. As Sam White reported in Lord Beaverbrook's

London Evening Standard, it left Rainier "the unchallenged master of all he surveys in his principality."[10] White proclaimed that, "It can be said that Mr. Onassis's power in the affairs of the casino is at an end. This power, based on his 42 percent shareholdings in the company, was exercised through an administrator appointed by him and primarily responsible to him. Now this administrator, M. Charles Simon, has resigned in circumstances which have given unconcealed satisfaction to both the palace and the French. Both have made it clear that never again can there be an Onassis man appointed to this key post." According to White, Rainier had summoned Onassis to the palace to complain that the Hotel de Paris rebuilding program was behind schedule and alarmingly over budget. He blamed Onassis's interference. "Mr. Onassis denied that he had interfered. Unfortunately, M. Simon was unable to support him in this denial and resigned in protest."

The tone of the story came as a shock to Onassis. The Beaverbrook newspapers had usually been kind toward him, and he had always regarded White, a drinking crony, a frequent dinner companion, and one of the paper's best reporters, as a journalist he could trust. That White had not called him for a quote before running the story made him suspicious as well as angry. To add insult to injury, White's column contained another item that was sure to offend Onassis and hurt his wife:

> INCIDENTAL INTELLIGENCE: Mrs. Tina Onassis is suffering from a malady that only the very rich can get, and is now rarely seen in Monte Carlo. She explains that life on the enormous Onassis yacht has produced a form of nervous breakdown.[11]

Max Beaverbrook often used his newspapers to settle personal scores, and it was no secret that he was among many of Churchill's friends who resented the way Onassis—who they viewed as a social upstart and an out-and-out crook—monopolized and exploited the old statesman's friendship. Onassis suspected that the item was the start of a Beaverbrook vendetta.

Ari knew there was little he could do about it himself. "Newspapers and widows," he was fond of saying, "always had the last word." Yet he knew that Beaverbrook, a notorious skirt-chaser whose loyalties, even at the age of eighty, could still be swayed by his randy ambitions, was "sweet on" Tina.[12] And although Onassis and Tina were now leading separate lives ("I think it's beastly that we still sleep together," one close friend recalls Tina, a little tipsy, saying at a dinner party, as if she were talking about other people, whose manners she found wanting[13]), Onassis persuaded his wife to appeal to the press baron to get off his back.

In a handwritten letter sent from Paris on April 15, she pleaded with Beaverbrook to ensure that in the future White "got the true facts before writing, if not for my sake for the sake of the newspaper." Her husband's relationship with the prince was most cordial, she insisted. Simon had been asked by Onassis to resign "for the simple reason he was getting fed up with him." While not denying White's story that she had suffered a kind of nervous breakdown, she added: "I have the greatest fun in my life when I am on the yacht, as you have seen yourself."[14]

Beaverbrook solemnly expressed his distress at her distress, and promised that his newspaper would always present *her* in a favorable light. "And indeed it would be impossible for the Paper to take any other course," he wrote chivalrously. But there was no mention of her husband; no promise was made to him. And although Beaverbrook wrote Tina that he was "far, far from London for ten months of the year . . . and the papers are not really under my control," such a man, Onassis knew, would never let his Machiavellian mind wither on the vine of retirement.[15]

When Randolph Churchill told Onassis that Joe Kennedy had rented a house in the south of France for the summer and was seeing a great deal of Beaverbrook, whose villa was not far away, Onassis jumped to the conclusion that the Kennedys were responsible for the more hostile tone toward him in Sam White's column.

The friendship between Beaverbrook and the Kennedys dated back to the time when Joe was the American ambassador in London. In 1956, Kennedy had endowed a Lord Beaverbrook chair at the University of

Notre Dame.[16] Now he wanted the favor returned: He asked Beaver-brook to arrange a meeting with Sir Winston, whose support for Jack's presidential run the following year would be useful. But Kennedy had been a vigorous advocate of appeasement toward the Nazis in 1940, and both Randolph Churchill and his father's private secretary, Anthony Montague Browne, were fiercely opposed to the meeting.

A former wartime RAF fighter pilot, Montague Browne pleaded with the old war leader "not to be seen consorting with an enemy of our country, who had sought to harm us in our hour of greatest need." Nevertheless, Churchill decided that "If Max (Beaverbrook) attaches importance to it, I will go. But it must not appear in the press."[17]

Meanwhile, Tina was growing uneasy. She demanded to know whether Onassis was having her followed. He wasn't, not at that particular time at least, and it worried him that someone was. "Who the hell would want to follow her apart from me?" he asked Gratsos.

Several weeks later, Tina again complained that she was being followed. "It's Beaverbrook's people—doing the Kennedys' dirty work for them," Onassis concluded vehemently.[18]

Although he had no evidence, at least none that would stand up in a court of law, the line that ran from the Kennedys to Beaverbrook—whose newspapers had suddenly and otherwise inexplicably turned on him*—was sufficient to deepen Onassis's paranoia about Bobby Kennedy's unforgiving animus.

---

* Anthony Montague Brown at this time was staying at Beaverbrook's villa in the south of France when Tina Onassis invited him on a short cruise to Corsica. Onassis was away on business, she explained, and she wanted some "young and jolly" friends to enliven the trip. Montague Browne cleared his brief absence with Churchill and, the most courteous of men, told Beaverbrook that he would be away for a few days and hoped he would be welcomed back at the villa when he returned. Although Montague Brown was well aware of his host's style, his love of intrigue, and the almost diabolic pleasure he took in his own skulduggery, Beaverbrook's reaction astonished him. "Ah yaas, I know all about that cruise," the press baron said mysteriously. "I'd rather you didn't go." When the Englishman protested, Beaverbrook told him: "Ah yaas, you'll be a most interesting co-respondent and my newspapers will duly report that fact" (Anthony Montague Browne, p. 195).

*   *   *

Onassis never trusted Sam White or any other Beaverbrook reporter again, and he lost the goodwill even of those who had usually treated him fairly, and often with genuine affection.

Unfortunately, this schism happened when Onassis needed all the media goodwill he could get—for he had fallen in love with the tyrannous prima donna Maria Callas.

FIVE

DANCING THE TANGO WITH ANOTHER PRIMA DONNA

Yes, I'm a murderer. I'm a thief. But
I am also a millionaire and powerful.
I will never give up Maria.

—ARISTOTLE ONASSIS

Onassis first met Maria Callas in September 1957 in Venice at a ball thrown by the American society gossip columnist Elsa Maxwell. "There was a natural curiosity there; after all, we were the two most famous Greeks in the world!" Onassis told his old Hollywood friend, Spyros Skouras.[1]

Nevertheless, although he admitted that he approached every woman as a potential conquest, two years passed before he made his next move, at Contessa Castelbarco's annual Venice Ball, in the spring of 1959. Dancing almost every dance together, it was apparent to everyone—especially to their spouses, Giovanni Battista "Tita" Meneghini and Tina—that he had embarked upon what Meneghini would later call Ari's "diabolical project" to seduce Callas.[2]

In June, Onassis followed the diva to London for the opening of *Medea* at Covent Garden. It was the London opera event of the year and he gave a candlelight supper party for her in the ballroom of the Dorchester Hotel to which he invited one hundred and seventy guests. When Callas said she regretted that nobody played tango music any more, he told the musicians to play nothing but tangos for the rest of the party.

Few there that evening could have doubted his intent as he propelled Callas around the floor to the erotic rhythms of the dance he had mastered like a gigolo thirty-six years earlier in Argentina.

Even in 1923 in Buenos Aires, Onassis knew that he was meant to be rich, to do big things. Even when he was just one of thousands of refugees pouring into the city from the eastern fringe of Europe—Greeks, Armenians, Syrians, Turks, Lebanese; all dismissed as Turkos—he already had a powerful sense of himself. He had a conspicuous nose, dark eyes, and polished black hair, and even though he was small, with a muscular torso and arms that almost gave the upper part of his body the appearance of deformity, he had a charm that women loved.

Onassis shared a room with a distant cousin and his wife in la Boca, the Italian quarter in the east end of the city. The room was above a dance hall; between the loud music and the frequent sound of passion in the next bed, sleep was difficult. The problem was solved when he got a night job on the switchboard of the telephone exchange and slept during the day.

Telephone traffic was slow in the small hours, and he worked on his Spanish, read the newspapers, and listened in on calls, especially business conversations. Discovering that the best deals were often done after the markets had closed, he was able to make five hundred dollars on a linseed oil deal, and another couple of hundred speculating in hides. He invested this in a few good suits, private tango lessons, and a subscription to a smart rowing club called l'Aviron (the Oar). He moved into a furnished room on Avenue Esmeralda.

It was the start of an extraordinary double life, an almost schizophrenic existence. His new friends at l'Aviron knew him only as a young man with money, business connections of an unspecified kind, and a good address. An instinctive performer, he was able to adapt his personality with extraordinary ease.

It was probably to create the illusion that he was a man of culture that Onassis went to his first opera. *La Bohème,* at the Teatro Colon. The Italian soprano Claudia Muzio, whose fame at its peak rivaled Caruso's, sang Mimi. At thirty-five, twelve years older than Onassis, she was a little past her prime, although she had a robust beauty and an appetite for young admirers.

Although sex was to play a key role in his life—his success, his fame, his ambition, all had their sources in a substratum of sex—Onassis had never taken the time to fall in love. Nevertheless, he sent Muzio flowers every day for a week. When he finally presented his card at the stage door, she invited him to her dressing room. "I expected a much older man," she told him. They became lovers.

Ari was doing okay. Muzio opened some of the social doors he wanted to enter, but he still hadn't got into the real money or discovered what he really wanted to do. He had worked in his father's tobacco business in Smyrna, before the Turks deported his family to Greece in 1922, but he did not want to follow his father's line. They had fallen out bitterly—over family money Onassis claimed to have spent on bribes to get his father out of prison—and Onassis was determined to prove he could make it on his own.

Nevertheless, when he overheard a Buenos Aires film distributor boasting on the phone to his New York office about the money the new Rudolph Valentino movie was making—"women can't get enough of this Sheik son-of-a-bitch"—Onassis wrote to his father in Athens, suggesting that they introduce Oriental tobacco to Argentina. The local cigarette manufacturers were using mostly Cuban tobacco, a dark leaf that women found too strong. He proposed using a milder brand, putting a sheik on the wrapper, and targeting the product at young women. His father shipped several bales of the finest leaf he had.

Onassis hired his cousins Costa and Nicos Konialidis, who had been orphaned in the Smyrna massacre and had followed him to Buenos Aires, to hand-roll his two brands: Primeros and Osman. Sometimes he would help out with the hand-rolling himself. But at 6:30, he always took a shower, changed into his best suit, did the rounds of the cafes, and visited l'Aviron. At 10:30, he switched clothes again and began his night-shift at the telephone exchange.

When his cigarette sales faltered, he turned to Muzio for help. Like most women at that time, she smoked privately but never in public. If she were to be seen smoking his cigarettes, it would be an act of emancipation, he suggested. "And it would help me a little bit, too, you know!"[3] It is not known what effect her endorsement had on sales, but she was the first of the many famous and influential figures he would use in his life to get what he wanted.

"I hate the opera," he admitted after their affair was over. "I think I must have a tin ear. No matter how hard I concentrate it still sounds like a bunch of Italian chefs screaming risotto recipes at each other."[4]

Now more than three decades later in London in 1959, he was dancing the tango with another prima donna. But if—on the edge of his sixtieth birthday, and driven by the imperatives of middle-age—he was again roused by the scent of conquest, Callas was no less impatient to surrender herself to him.

Onassis felt no pity for the husband he was planning to cuckold. He had an intuitive sense about other people's sex lives, and told Johnny Meyer: "Meneghini's not servicing the account. What does he expect?"[5] By the time Callas's London engagement concluded at the end of June, Ari had become her lover.

It was almost certainly the first time she had slept with a man other than her husband, according to Franco Zeffirelli, her favorite director and friend. After her seduction, Callas appeared to lose her nerve. Her first reaction, says Zeffirelli, was to "cling to her 'sacred' marriage."[6]

She certainly made a show of avoiding Onassis when he telephoned her at her Italian villa in Sirmione on the shores of Lake Garda to invite her and Meneghini to join him on a cruise aboard the *Christina*. But her reluctance to take her lover's calls was probably only a performance to disarm her husband. Her dressmaker, Madame Biki, was already preparing a cruise wardrobe: twenty dresses, trouser suits, bathing costumes, and negligees.[7] Curiously, Meneghini seemed to suspect nothing. "This invitation comes at just the right time. The doctor recommended sea air. They say the Greek's yacht is very comfortable. Let's give it a try. If you don't like it, at the first port we can return home," he told his wife.

On July 21, the Meneghinis flew to Monte Carlo. They dined at the Hotel de Paris with Onassis, Tina, and Elsa Maxwell. Maxwell had not been invited on the cruise, which was to include Sir Winston Churchill and his wife, Clementine; their eldest daughter, Diana Sandys; Churchill's doctor, Lord Moran; Anthony and Nonie Montague Browne; and Onassis's sister Artemis and her husband, Dr. Theodore Garofalides. Being excluded deeply wounded the old society fixer, who felt that Onassis and Callas owed her a favor for having brought them together at the Venice ball. In an insidious valedictory note, delivered on the eve of the cruise, Maxwell reminded Callas that she was "taking the place of Garbo [formerly a regular Onassis guest and rumored lover], now too old, on board the *Christina*." From this moment, Maria must "*take* everything,"[8] Maxwell slyly egged her on. ("That gossip already knew everything and was meddling with the slobbering advice of a bird of ill omen," Meneghini complained in his memoirs.)

Onassis relished the sexual competitiveness that yachts and beautiful women engender, and life aboard the *Christina* had an innate undercurrent of "jealousy and intrigue—it was part of its excitement," said a frequent woman guest. Meneghini quickly picked up on this, although his attempted games of footsie beneath the table with Nonie and other women guests tended, as Nonie's husband drily explained, "to leave their slacks or skirts covered in white powder (from his deck shoes) as evidence of his attempts to establish a closer acquaintance."[9]

But not even to the eyes of those versed in the aesthetics of conspiracy did Callas behave like a woman who was having an affair with her host. If recent deceits, and the deceits still to come, troubled her conscience, she did not show it. Montague Browne, with his diplomat's eye for human foibles, thought that she almost seemed to be trying to parody the stereotypical prima donna in her behavior.

Meneghini wanted to leave the cruise at Capri; Callas refused. Cut off by language (he spoke no English and very poor French), temperament, and a queasy stomach, he spent more and more time in their state-room, his resentment simmering. He sneered at Onassis's dark physique ("he didn't seem to be a man, but a gorilla"[10]) and unkindly compared Callas's body to Tina's. The strain in the Onassis marriage also began to reveal itself. Appearing in a clinging scarlet silk dress, Tina asked Montague Browne coquettishly, "Do you like my new dress, Anthony?"

"Wow! Yes indeed!" he told her, thinking it might have been painted on her.

"I think she looks like a little French tart," Onassis growled angrily rolling his r's, a habit he tried to restrain in polite society.[11]

Inevitably, the strain in the two marriages quickly became apparent. "It was sad to see the (Meneghini) marriage disintegrating, the tenderness fade," Nonie Montague Browne wrote in her diary. "Tina continued to be a sparkling hostess, but one was aware of the well controlled tension."[12]

By the time they got to Rhodes, Onassis was drinking heavily, and "began to pay marked attention to Maria Callas."[13] They had not made love since the end of June in London when, according to Onassis, she had performed the act of fellatio on him in the back of his Rolls-Royce as they drove down Park Lane, and perhaps the abstinence was beginning to tell on both of them.

On August 4, the *Christina* dropped anchor at the old port of Smyrna. It was an important landmark in Onassis's life. Reflected through the prism of his fame and fortune, the story of his youth would always excite and impress his guests, especially the women, as he led them through the narrow streets in which he had played as a child and described how it had been before the Turks came in 1922, before the holocaust; and as he took them to where his father's warehouses had been, and to the small graveyard in which his mother was buried.

Like all good storytellers, he was not content with the mere recital of facts. He would recall the eerie silence that filled the streets as the advance guard of the Fourth Turkish Cavalry regiment, dressed in black and carrying drawn scimitars, entered Smyrna in the early morning; the first night of the holocaust, the flames appearing, as fire does at night, to plunge from above, covering the city with molten cinders; his grand-mother Gethsemane reading from Ecclesiastes ("To every thing there is a season, and a time to every purpose under the heaven") in Turkish, the only language she knew, as the family waited for the enemy to arrive at their door.

After Churchill had gone to bed, Onassis took Callas and Meneghini to visit the bars and brothels where forty years earlier in bedrooms of red-lined silk, local girls had begun his education in the world of physical love and instilled in him a view of women ("One way or another, sweets, all ladies do it for the money") that would destroy every love affair he would ever have.

According to Meneghini, they spent the night "in the company of dealers, prostitutes, and assorted sinister characters."[14] It was five o'clock in the morning before he persuaded his wife and her lover to return to the *Christina*. Onassis was drunk. "He couldn't stand up, he couldn't talk," crowed Meneghini.[15] It may have been one of the last truly happy moments of his life.

At Istanbul, once the heart of Byzantium, the soul of the Ottoman Empire, the Patriarch of Constantinople and Head of the Greek Orthodox Church was invited on board for lunch and a short cruise up to the Black Sea and back. Onassis knew that sailors believe a priest on board brings bad luck, and he anxiously told his captain: "This guy is one hundred priests' value and there is an eight knot current running, so be very, very careful."[16]

Impressed by "the world's finest singer and the most famous mariner of the modern world, a modern Ulysses,"[17] the Patriarch bestowed blessings upon the lovers as they knelt side by side on the deck, and offered prayers for the honors they had brought to their homeland. It seemed to the wretched Meneghini as if a marriage rite were being performed.

Few on board were unaware of Callas's rendezvous in the middle of the night with her lover. These trysts had become "more obvious and indeed obsessive," according to Montague Browne, who nevertheless politely preferred to imagine they were meeting simply to share *Saletes Grecques* (a snack) in the ship's galley.[18]

On August 9, the *Christina* dropped anchor at Athens, and Onassis's sister Artemis and her husband Theodore gave a party at their home in Glyfada for Onassis's old Hollywood friend, Spyros Skouras. The lovers barely left each other's side. On her wrist Callas wore a single gold bracelet. Inscribed on the inside were the initials *TMWL: To Maria With Love.*

"Ari is such a creature of habit," Tina told Artemis when she saw the telltale gift on Callas's wrist.[19]

Tina Livanos had first met Onassis in her father's suite at the Plaza Hotel in the spring of 1943. She had broken her leg in a riding fall and had hobbled into the room on crutches. She was fourteen years old. Afraid to admit how strongly attracted he was to her, Onassis simply wrote in his notebook: "Saturday, April 17, 1943. 7 P.M." He was, he later admitted, too nervous to write her name. But he also knew that the first rule in keeping secrets is to leave nothing on paper.

Born a British subject of Greek lineage, she did not look remotely Greek. She had become a United States citizen by an act of Congress during World War II, and now her clothes, her gestures, even her humor were extraordinarily American. Only her voice remained stunningly English.

She knew what she was in for when she surrendered to Onassis's charms. ("Anthony," she would later ask Anthony Montague Browne, "did you know that Ari seduced me in a Chriscraft on Long Island Sound when I was sixteen?") Onassis was cut from the same cloth as her father; if she married him, she would be delivering herself from one master to another. Nevertheless, like all Greek girls of her class, she had been bred for the kind of marriage that makes alliances of powerful families. She also wanted excitement in her life, and Onassis seemed very foreign to her English-educated mind, unlike Niarchos, who had also expressed a deep interest in her, despite the fact that he was already married. Listening to Ari talk about his past reminded her of Oriental storytellers in the marketplace. She never knew what to believe, what to dismiss; the mystery at his core excited her girlish imagination.[20] "All anybody will ever really know about Ari's early days," she told her sister Eugenie, "will be what he chooses to tell them."

No man had treated her as a woman before, and she was flattered as well as fascinated by him. But when he asked for her hand, her father was appalled. This was not because he knew about Onassis's chronic unfaithfulness, his heavy drinking, and his dangerously violent temper. Cynics said that the prospect of a Livanos–Onassis shipping alliance finally outweighed the risk of exposing his daughter to such a man. What angered him was that Onassis had asked for the wrong daughter! Daughters marry strictly in order of seniority, he told Onassis, and Eugenie was the first in line. "Your daughters aren't ships, Mr. Livanos, you don't dispose of the first of the line first," Onassis shot back.[21] "A man must observe the rules in life," Livanos told him. "The rules are there are no rules," Onassis said.[22]

Nine months later, on December 28, 1946, Onassis and Tina were married at the Greek Orthodox Cathedral in New York City. She was thrilled on her wedding night when he slipped a gold bracelet on her wrist inscribed *Saturday, April 17, 1943, 7 P.M—T.I.L.Y. [Tina I Love You]*.

Perhaps it was these memories that tore at Tina's heart now.

Unable to sleep, she had gone up on deck for some fresh air in the early hours of August 12, 1959, when she saw her husband and Callas making love in the saloon. She woke Meneghini and told him what was happening. But as she would later tell a lover, she "could see by his face that he was beyond doing a thing about it."

She had accepted that her own marriage was over, and what her husband got up to no longer mattered to her. However, in his memoirs, written a decade later, Meneghini remembered it differently: Tina was "almost completely naked" when she entered his cabin and threw herself sobbing on his bed: His hint that she had offered herself to him in an act of mutual commiseration seems improbable, yet stranger liaisons had happened in the past—and would happen in the future—aboard the *Christina*.

*   *   *

On August 13, the *Christina* returned to Monte Carlo, and Callas and Meneghini flew back to Milan. It was not long before Callas began musing to Meneghini about a separation: "I sometimes wonder how different our lives would be if we were no longer together."

The European newspapers were soon filled with stories about Onassis's obsession with the prima donna. In Venice, Tina was reported dancing cheek to cheek with Count Brando d'Adda. "Does her heart belong to d'Adda?" Johnny Meyer cabled a friend. The four-way split titillated society with its promise of scandal and tears.

Nevertheless, Onassis and Meneghini at first tried to contain the damage the rumors were doing to Callas's reputation. Callas, who relished watching her lover in action, insisted on being present when the two men debated her future. She loved the look of ruffian strength in Onassis's body; his power and his sexuality were indivisible. The idea that he was fighting with her husband for her favors aroused her deeply, and one dinner at Sirmione was particularly exhilarating.

Onassis had been drinking whisky steadily during the drive from Milan, and was as ecstatic at the thought of taking Callas from her husband as she was at being possessed by him. She had dressed for dinner in a strapless black chiffon gown. Onassis wore his familiar dark glasses and a dinner jacket. In spite of the warm summer evening, Callas insisted that a fire be lighted in the dining-room fireplace. Sweating like a burro, Meneghini lost first his appetite, then his temper. But more than anger, he was driven by a desire to put up a final fight that he knew he would have to live with for the rest of his life.

When Onassis taunted him, and with licentious glee demanded to know how many millions he wanted for his wife: "Five, ten?" Meneghini told him: "Stick your money up your Greek ass." (In his memoirs, the riposte loses something in translation: "I replied, 'You are a poor drunk and you turn my stomach. I would like to smash your face in but I won't touch you because you can't even manage to stand up.' ")

According to Meneghini, Onassis began to weep and made an emotional and extraordinary statement:

"Yes, I'm a disgrace. I'm a murderer. I'm a thief . . . But I am also a millionaire and powerful. I will never give up Maria, and I will take her away from whomever it's necessary, using whatever means, sending people, things, contracts, and conventions to hell."

Callas began screaming hysterically, according to Meneghini, who told her lover: "I am placing a curse on you that you never have peace for the rest of your days." [23]

It was a remark he would look back on with a sense of satisfaction.

SIX

THE PRINCE, THE WIFE, AND HER LOVER

Ears had they, and heard not.

—AESCHYLUS

Battista Meneghini may not have been the only person who believed that Onassis was capable of killing anyone who got in his way, but he had more reason than most to fear him.

A Verona lawyer, one of the intermediaries Callas used to try to persuade him to concede her a divorce, recalls his story of Onassis's murder confession. "I said, 'Battista, if Onassis had murdered somebody, I don't think he would brag about it to you, or to anyone else.' He said, I remember his words exactly: *'Al bugiardo non si crede la verita.'* (Nobody believes a liar, even when he is telling the truth.) I thought, poor Battista. A hungry ass will eat any straw." [1]

Shortly after this conversation, Callas herself threatened to shoot Meneghini. "Be careful Battista; one day or another I'm going to arrive at Sirmione with a revolver and I'm going to kill you," she threatened, according to Meneghini's memoirs. [2]

"That wasn't Maria talking. That was Onassis," he told friends. Onassis would always get others to do his dirty work. "You have no idea what he is capable of. When he is finished with Maria he will toss her overboard without a second thought," he said.

Franco Zeffirelli, who had no time for Meneghini, nevertheless shared his fears for Callas's safety. "One had heard rumors of how these Greek tycoons treated their unwanted women—stories of violence and even murder," he says.[3]

In September 1959, Callas announced that her marriage was over. The break had been in the air for some time, she said; her cruise aboard the *Christina* had had nothing to do with it.

Her affair with Onassis had made her famous in a way she had never been as an opera star. She loved the attention every bit as much as Onassis did. Tracked down at Harry's Bar in Venice, Onassis admitted: "How could I help but be flattered if a woman with the class of Maria Callas falls in love with someone like me? Who wouldn't be?"[4]

But as one Callas biographer would shrewdly observe, "The little Smyrnan refugee with his tankers and his billions needed these shots of flattery like an addict, and with time and his advancing age, he needed bigger and bigger doses for the same effect."[5]

In October, Meneghini filed suit for a legal separation. His writ did not name Onassis, referring only to Callas's abrupt transformation from a "loyal and grateful wife" to one whose behavior was "incompatible with elementary decency" following a cruise with "persons who are reckoned the most powerful of our time." And on November 14, after a six-hour settlement hearing (Maria got the Milan townhouse, most of her jewelry, and the income from her recording royalties; Meneghini kept their house at Sirmione and all their real estate investments), the separation was made legal.

Eleven days later, Tina Onassis filed for divorce in New York State Supreme Court on the grounds of adultery.

On the morning the story broke in the New York papers, Onassis received a small package containing a gold bracelet inscribed *Saturday, April 17, 1943. 7 P.M. T.I.L.Y.*

*   *   *

Onassis's statement to the press was terse: "I have just heard that my wife has begun divorce proceedings. I am not surprised, the situation has been moving rapidly. But I was not warned. Obviously I shall have to do what she wants and make suitable arrangements."

Now approaching her thirty-sixth birthday, Callas had expected much more—a declaration of his love for her, a public commitment to their future.

But Tina's announcement had deeply shocked Onassis. He was not the first man, nor would he be the last, who wanted both his mistress and his wife. In London, he broke down in Spyros Skouras's suite at Claridge's and "sobbed like a child" after pleading with Tina to change her mind for more than an hour on the telephone.[6]

He was still begging her to return when they met in Paris in April to discuss the terms of the divorce. By setting them up to be photographed dining together, he encouraged speculation of a reconciliation. But there was no relaxation of that wariness with which warring couples treat each other ("Even in love," Tina would later write a friend, "he could not give himself without bargaining"[7]), and there was no new beginning.

To please her father, Tina dropped her allegations of adultery, and in June got an uncontested "quickie" divorce in Alabama on the grounds of mental cruelty. Turning the screw, she reverted to her maiden name, Livanos, which, she said, had a cachet that was considerably smarter than her former husband's name. Friends were shocked at her bitterness. "Even if she thought that Ari's name was shit, it was still the name of her children," Theodore Garofalides said sadly.[8]

In retaliation, Onassis went out of his way to be seen celebrating his new freedom with his mistress. There were pictures of them dining together in London and Paris, clubbing in the south of France, dancing together in Rome. "It is impossible for them to dance cheek to cheek as Miss Callas is slightly taller than Mr. Onassis," ran a caption to a picture of them on the dance floor. "But as they dance she has lowered her head to nibble his ear . . ." Speculation that marriage was imminent was rife. At last Callas could see their future together.

*   *   *

Spyros Skouras, who had just bought the movie rights to Bobby Kennedy's recent bestseller, *The Enemy Within*—about Kennedy's time as counsel for the Senate Rackets Committee, and his war with labor boss Jimmy Hoffa*—told Onassis that John F. Kennedy planned to run for president in 1960.† Whether this was an informed guess or inside information from Bobby is anyone's guess. Either way, Skouras was right; in December, he urged Onassis to mend his bridges with Bobby; if his brother won in November, he could become "somebody you don't want as an enemy."[9]

Skouras liked Onassis, despite the fact that they had almost come to blows when Onassis had refused to contribute to his Greek War Relief fund in World War II. ("I've seen what happens to men who get involved in political causes,"‡ Onassis had told him, and Skouras, a proud Peloponnesian Greek, had called him "a jumped-up shit from Smyrna."[10]) Skouras said that if he and Onassis had been able to bury the hatchet, so should Onassis and Bobby.

But it was not as simple as that.

Bobby Kennedy's animus toward the corrupt and powerful Hoffa was easy to understand. The antagonism Bobby felt for Onassis was personal, more complex, and infinitely more difficult to explain.

Franklin D. Roosevelt, Jr., who had known Onassis since 1942, when they met at a birthday party on Long Island for Onassis's Norwegian mis-

---

* Ironically, according to Budd Schulberg, who was to write the screenplay, the film eventually fell through because "the large studios were scared of offending people who are the fairly powerful force in this country—the big labor racketeers . . . It was just simple old-fashioned fear." (Schulberg, cited in Jean Stein, *American Journey: The Times of Robert Kennedy* [George Plimpton, ed.] p. 58. London: Andre Deutsch, 1971.)

† His candidacy was officially announced on January 2, 1960.

‡ In 1922, Onassis's favorite uncle, Alexander, who had become involved in the Asia Minor Defense League, a separatist movement seeking an autonomous Greek zone inside Turkey, with Smyrna—with its preponderantly Greek population, and as the traditional centre of Christianity in Asia Minor—treated as an international zone, had been found guilty by a Turkish military tribunal and hanged in the public square of Kasaba.

tress Ingeborg Dedichen, believed that Onassis exaggerated Bobby's animosity toward him. "Bobby doesn't like him, but he's far too busy to pay Ari as much attention as Ari pays him," Roosevelt told Costa Gratsos in 1960.[11]

Shortly before the Justice Department indictment against him for his war-surplus tanker scam was unsealed in 1954, Onassis had asked Roosevelt to represent him. Roosevelt passed. About to make his bid for governor of New York, he had feared that the scandal that attached itself to Onassis's name would damage his chances.*

Perhaps another reason Roosevelt passed up the big bucks that Onassis had offered him was Ari's insistence that Bobby was the architect of his misfortune. "If I can't stare down a little runt like Bobby, what would be the sense of taking on the Justice Department?", Onassis had said. This attempt to assure Roosevelt of his determination not only to take on the U.S. government but also to smear Kennedy in the bargain was worrisome.[12] The Kennedys had been friends of the Roosevelts for years; Joe Kennedy had help finance his father's first presidential campaign in 1932; five years later, FDR appointed Joe Kennedy ambassador to the Court of St. James.

Although a man of considerable *amour-propre*, Roosevelt had never been known for his political smarts. But in 1960, as he anticipated hitching his star to the rising JFK's—as Joe Kennedy had once hitched his to FDR's—he must have felt that he had finally gotten something incredibly right in his life.

---

* Far more damaging to Roosevelt's prospects was his own deeply flawed character. His name and physical similarity to his famous father gave him a cachet that many believed his abilities failed to justify. He had a poor record in the House, which earned "the contempt of colleagues and of many constituents alike for his laziness and absenteeism. Equally damaging, he had not played ball with party bosses—partly out of principle but also because of his assumption that his name put him beyond their reach." In the event, the Tammany bosses steered the governor's nomination away from Roosevelt to Averell Harriman. Although Roosevelt would run for office again and again in the future, each time he would finish "further away from the victory that had once seemed assured." (Peter Collier with David Horowitz, *The Roosevelts: An American Saga,* p. 461. London: Andre Deutsch, 1994.)

*   *   *

Meanwhile, marriage to Onassis remained elusive for Callas. "Anthony, tell Ari he ought to marry me," she pleaded with Montague Browne one evening at dinner. "Maria, I can't do that," Onassis interrupted sharply. "This is a pay-as-you-go arrangement."[13]

But when friends pressed her to leave him, she said: "When slight has followed slight, and insult has been added to insult, the love which remains is often illogical, but it is also indestructible."

It was a kind of madness, she said poignantly, and nobody chose to be mad.

Her relationship with Onassis had not brought her the happiness she yearned for, nor the peace she deserved. "Whenever I saw them together, she would try to disguise [her unhappiness] and she obviously still doted on him," says Zeffirelli. "In an unpleasant way, this desperate adoration on her part only brought out the worst in him, tempting him to treat her as badly as he could. There seemed not to be any limit to his sadism."[14]

As her fortieth birthday came into view, and she recognized that both professionally and privately she was no longer in her best years, she began to suspect that Onassis had never intended their affair to be more than an adventure sanctioned by Meneghini's weakness and Tina's license.

Yet she also knew that if it were to start all over again, she would act in exactly the same way.

Meanwhile, still basking in her glory, Onassis never stopped trying to mold her, to demonstrate his power over her. He made her wear black, his favorite color. As if he were ordering some ritual of possession, he told her Paris hairdresser to cut off her big hair. Her new shorn look took ten years off her age, but it also deprived her face of its elemental excitement.

Callas often went for months without practicing a note. Onassis wanted her to go into the movies, and it was to this end that Franco Zeffirelli visited Skorpios to discuss a film production of *Tosca*. But their first dinner together was a disaster. As the evening progressed, the tension between Onassis and Callas became increasingly evident. Zeffirelli began to suspect that Onassis's interest in the project was another of "his vicious tricks: to lead Maria on and then let her down." [15]

Eventually, Callas rushed to her room in tears.

"I am," she told Zeffirelli pitifully when he went to her room to comfort her, "quite simply at his mercy." [16]

A few weeks later Zeffirelli was astonished when "a rather shady-looking Greek" called to see him in Italy and handed him a bag containing $10,000 in cash—Onassis's promised contribution in development money for the movie.*

It was "all very hush-hush, no receipt, nothing," [17] recalls Zeffirelli.

Onassis was a man to whom intrigue was as nourishment, of course. But this was a precursor of a deal that would eventually shock the world.

On January 20, 1961—the eve of John F. Kennedy's inauguration as president of the United States—Onassis celebrated what he claimed to be his fifty-fifth birthday in Monte Carlo. (Still pretending to be six years younger, the deception was increasingly difficult to sustain as the years passed, and his hard drinking and high lifestyle took their toll.) He had reached an age, he told the guests gathered to celebrate the occasion at the Hotel de Paris, when a man should take stock of his life, and he had decided to shift down a gear from the pace of earlier times.

Among the new crowd surrounding him—most of them very rich or very grand, which he knew were not always the same thing—were Prince and Princess Stanislas Radziwill. It was not the exiled Polish prince's defunct title that impressed Onassis so much; Radziwill had become plain mister when he took British citizenship in 1951. Nor was it his ready supply of amphetamines from Max Jacobson, the New York soci-

---

* This was a sign of how Onassis viewed the project, Zeffirelli would later reflect. "The pre-production costs of a film run to hundreds of thousands of dollars; by giving so little, he could appear to be promoting Maria, while actually ensuring that nothing happened." But after they failed to acquire the film rights to *Tosca*, Callas called Zeffirelli and demanded her money back. When he informed her that it had already been spent and, in any case, it had been Onassis's money, she told him: "It was my money. He made me pay it out of the little I have left."

ety physician known as "Dr. Feelgood."* Instead, it was the fact that his wife, Lee, was now the sister-in-law of the most powerful man in the world; "the guy I didn't ask to stay to dinner," Ari recalled ruefully.[18]

"But you *hate* the Kennedys," Gratsos reminded him when Onassis anticipated a closer relationship with Lee.

"Just Bobby," Onassis replied, adjusting his aim.[19]

Lee Radziwill and her husband had been married for only nine months, but she had already been drifting apart from her second husband when they got caught up in the Onassis circus. At twenty-nine, an Eastern Seaboard post-debutante (Farmington, three terms at Sarah Lawrence), with a nice smile and great hair, Lee was a lightning rod for men like Onassis. He admired "the way she was at home with wealth, the way she took luxury in her stride."[20] She had been born with a taste for money—Onassis believed that there was no higher kind of *chic*—and their relationship swiftly and inevitably became close.

If it were true that Bobby Kennedy had "an immense sense of humor about sinners,"[21] it definitely did not extend to Onassis. Bobby knew that Ari's affair with Lee would not remain private for long—Onassis himself would see to that—and Bobby clearly regarded the consequences with trepidation.

Briefly, the problem was this: Lee had married her first husband, Michael Canfield, in the Catholic church, and the Vatican refused to accept the validity of her divorce. Much time and Kennedy money had been lavished on trying to get the Canfield marriage annulled so that she could marry Stanislas—known as "Stas" (pronounced "Stash")—Radziwill, in the bosom of the Roman Catholic Church. Unfortunately, the wheels of ecclesiastical courts grind slowly, much too slowly for Lee who, in love—and

---

* It was Radziwill who introduced Jacobson to Charles Spalding ("I'd see Stas jumping around town and went to see Max," he told Seymour Hersh, *The Dark Side of Camelot*, p. 235 [Boston: Little, Brown, & Co., 1997]), and Spalding who introduced Jack Kennedy to Jacobson. In an unpublished memoir, Jacobson wrote that he first treated Kennedy with a shot shortly before one of the televised debates with Richard Nixon in the fall of 1960.

also pregnant—had gone ahead and married her prince in a civil ceremony in Virginia. Incensed at her impatience, the Holy Office in Rome ruled the Canfield marriage valid. When Kennedy became the first Roman Catholic president of the United States, the family felt that the Vatican should think again, and Bobby took charge of the campaign. "You supply the prayers, we'll provide the pressure, and God will deliver the miracle," he had told the Radziwills.[22]

It was at this point—as His Holiness was obligingly dematerializing her first marriage to Canfield—that Lee had fluttered into her affair with Onassis. Unfortunately for the Kennedys, Stas cared as little about Lee's affair as she minded his; by this time, he had focused *his* amorous interest and most of his financial hopes on automobile heiress Charlotte Ford.

Behind Stas's man-of-the-world air and dapper Adophe Menjou mustache was the insecurity of a stateless aristocrat hanging on for dear life to a token title that could still be traded in for directorships and invitations to the right parties. He had made, it was true, a few killings in London real estate, but Lee's addictions—travel, fashion, the high life, and keeping up with Jackie—were expensive items in the Radziwill accounts, and Stas, who was nineteen years older than his wife, had become increasingly querulous about the mounting cost of her aspirations.

According to David Metcalfe, one of Stas's oldest friends in London, "Lee's lovers were often men Stas was doing business with."[23] It came as no surprise to their friends, therefore, when Stas was rewarded with a lucrative directorship of Olympic Airways soon after Lee began her affair with Onassis.

His enemies unkindly dubbed him Pimp Radziwill. For Bobby, this was the final straw. He told Stas that he had to put a stop to Lee's affair with Onassis at once.

But Radziwill refused to acknowledge that there was a problem. And while Bobby fumed, the three of them—the prince, the wife, and her lover—continued to pursue their own agendas with the eager synergy of the seven dwarfs digging away in their diamond mine.

SEVEN

HAPPY BIRTHDAY, MR. PRESIDENT

Something's Got to Give

—TITLE OF MARILYN MONROE'S LAST FILM (UNFINISHED)

I n May 1962, Onassis lunched with his old friend Spyros Skouras at "21" in New York. He was fond of the tough, barrel-chested Greek who had emigrated to America in 1912 and became one of the "marvelous monsters" who invented Hollywood. In 1935, he had masterminded the merger of the old William Fox Corporation with Twentieth Century and created one of the most famous movie companies in the world. But at the age of seventy-three, things had started to go badly wrong for him and his studio.

The Elizabeth Taylor–Richard Burton epic, *Cleopatra,* which was being filmed in Rome, was beset by the scandal of their affair and the breakup of their marriages. The problems of Taylor's unreliable health and Burton's heavy drinking made headlines around the world. That spring, when the picture entered its thirty-third month of production, its budget had risen to the then unheard-of sum of $30 million.* Another

---

* In contemporary figures, $435 million.

movie—*Something's Got to Give,* starring Marilyn Monroe—was also running into difficulties because of the actress's chronic absenteeism. In thirty-five days of filming, she had turned up only a dozen times, and when she did arrive she was so heavily medicated she could barely remember her lines. With the company hemorrhaging money in Italy, the studio appeared to have lost control of its stars.

Skouras's days at the studio were obviously numbered, and like so many sentimental Greeks he wanted to become a shipowner in his old age. He already owned two tankers and, planning to build a fleet of modern container ships, wanted Onassis's advice.

According to Yannis Georgakis, the lunch became a "black farce of mutual paranoia, self-absorption, crossed lines and fate."[1] Nevertheless, Onassis listened patiently as Skouras unleashed his anger at those he felt had let him down: Taylor, Burton, *Cleopatra's* director Joe Mankiewicz, Marilyn Monroe. But when Skouras suddenly turned his venom on Bobby Kennedy, Onassis sat up with a start. How could Bobby Kennedy be responsible for his troubles? How? Because he was sleeping with Monroe and filling her head with ideas about how much the studio undervalued her, that's how, Skouras told him bitterly.

Time had only hardened Onassis's conviction that Bobby was at the bottom of his own problems; and the discovery that the president—who had been sleeping with Monroe on and off since 1954 when they met at one of Charlie Feldman's parties—had passed her on to Bobby was as exciting as an unheralded windfall. But was Skouras sure it was Bobby Kennedy, the attorney general, who was sleeping with Monroe? Skouras told him, "that goddam was'e aspace sonabitch with his Bible in one hand, and his preek in the other, of course I'm sure it's Bobby. He's worse than his sonabitch old man."*

---

* Skouras's antipathy to Joseph Kennedy went back to 1940 when he had attended a private meeting of film men at which Kennedy had urged them to stop making anti-Nazi movies and using their pictures to "show sympathy to the cause of the democracies versus the dictators," according to a letter Douglas Fairbanks, Jr., wrote to President Roosevelt in 1940. (Nigel Hamilton Papers, Massachusetts History Society.)

Skouras's animosity toward Bobby was probably exacerbated by a skirmish a few weeks earlier, when the attorney general had asked him to arrange for six of his friends to visit the *Cleopatra* set in Rome and be entertained at lunch by Elizabeth Taylor. Furious at his insensitivity, Skouras had wired back, "*Cleopatra* isn't a social outing but business."[2] The following morning, Bobby had called Skouras at his home in Rye, New York. "He told me, 'You Greek sonabitch, I make you wish you never come to America,' " said Skouras.[3]

There may also have been an element of sexual jealousy in Skouras's rancor, for he too had been Monroe's lover, and still spoke warmly if somewhat incestuously of his "won'erful Mahlin [who is] like my own daughter, hones'-to-Gah."[4]

But according to Arthur Miller, Monroe's third and final husband, Skouras himself was at least partly to blame for the Monroe crisis that was helping to cripple the studio. "She was furious at his denying her the ordinary perquisites of a great star," Miller says.[5] And Rupert Allan, Monroe's publicist and one of her last true Hollywood friends, believed that Skouras had deliberately tried to shift the blame for her costly absences onto Bobby Kennedy. "The mistake people made about Skouras was in thinking that he was just a thug. The truth is, he was a Machiavellian thug," said Allan, who suspected that Skouras was settling a private score with Kennedy when he refused to allow Monroe to appear at the President's forty-fifth birthday gala at Madison Square Garden in New York.[6]

Despite Skouras's attempts to prevent her going, a helicopter landed on the Fox lot shortly after noon on Thursday, May 17, and Kennedy brother-in-law Peter Lawford, acting like the D-Day hero he had played in the studio's *The Longest Day,* swept Monroe off to Los Angeles International en route to New York to perform her notoriously orgasmic rendition of "Happy Birthday" for the president.

Thus, events were set in motion whose consequences would inextricably entwine the lives of Bobby, Jackie, and Aristotle Onassis forever.

When Bobby Kennedy failed to dissuade Lee Radziwill from seeing "the Greek," he telephoned Ari in Paris and demanded that Onassis stop seeing Lee. Onassis later dined out on his riposte: "Bobby, you and Jack fuck your movie queen and I'll fuck my princess."

Shortly afterward, the Kennedys cut Monroe loose.

Monroe's grip on reality was indeed shaky if she had persuaded herself that Bobby really would divorce the mother of his seven children. As the attorney general of the United States, a member of the most famous Catholic family in the land, and a politician who had just been named Father of the Year (an award that celebrated his "Norman Rockwell family life"), he was unlikely to run off with a thrice-married Hollywood sex symbol. Nevertheless, when the penny did drop, she felt abused. "She was heartbroken . . . she felt the Kennedys had passed her around like a piece of meat," said Rupert Allan.[7]

Onassis was furious at the way the Kennedys had treated her—and how, as he saw it, they had used her to help destroy his old friend, Spyros Skouras. The day Monroe's picture *Something's Got to Give* was closed down and she was fired by the studio, Onassis called Allan, whom he had known since Allan's days as Princess Grace's press agent in Monaco. "Ari told me to tell Marilyn that his yacht was at her disposal for as long as she wanted it," Allan recalled. "He said, 'She offered to do me a great favor once, and I owe her.'"[8]

Monroe had always known how to stage a scene to get what she wanted, and she threatened the Kennedys. "If I don't hear from Bobby Kennedy . . . I'm going to call a press conference, and blow the lid off this whole damn thing . . . I'm going to tell about my relationships with both Kennedy brothers," she told Robert Slatzer, a former lover.[9]

Bobby flew to California to prevail upon her to go quietly. Their affair—her affair with *all* the Kennedys—was over. But ending a love affair is rarely easy, and Bobby bungled it.

There are contradictory accounts of what happened between about one o'clock on Saturday afternoon, August 4, when Bobby Kennedy met Monroe for the last time at her house in Brentwood, and around 10:00 P.M. when the actress was found dead in her bedroom by her housekeeper, and 4:25 the following morning when the police were finally called—after Bobby had been helicoptered out of town.

The autopsy report certified the cause of death as "acute barbiturate poisoning due to ingestion of overdose," but the manner of her death would never be convincingly established. She may have died by her own hand—or by someone else's. But such a massive cover-up operation had been performed by Peter Lawford, and others, in those five or six hours between the discovery of her body and the call to the police, that it would be impossible to establish anything beyond the fact that Marilyn Monroe's life was over, and the legend had begun.*

Nobody had more powerful instincts of self-preservation than the Kennedys. And fleeing from the dark morass of the Monroe tragedy, Jack and Bobby swiftly set about validating their credentials as family men: By that autumn both their wives were dutifully pregnant, Ethel for the eighth time, Jackie for the fourth.

The announcement that the First Lady was pregnant was as carefully timed as anything John F. Kennedy had done in his whole political life. Conscious of his narrow victory in 1960, when he had won by a margin of 114,000 votes out of 68 million cast, a newborn baby in the White House for the first time in the century would be a public relations godsend in the run-up to the 1964 poll.†

*   *   *

---

* Natalie Jacobs was with her future husband Arthur Jacobs, Monroe's publicist, at a Henry Mancini concert at the Hollywood Bowl on the evening of August 4, when the news reached them by 11:00 P.M. that Monroe was dead. "We left the concert at once and Arthur left me at our house. He went to Marilyn's house . . . My husband fudged everything." (Anthony Summers, *Goddess: The Secret Lives of Marilyn Monroe,* p. 461. London: Warner Books, 1992). According to Peter Lawford's third wife, Deborah Gould, Lawford "went (to Monroe's house) and tidied up the place, and did what he could, before the police arrived . . . That's where Peter's role came in, to cover up all the dirty work, and take care of everything (for the Kennedys)" (Summers, p. 465).

† The last U.S. President to become a father while in office was the Democrat Grover Cleveland, whose daughter Marian was born in the White House in 1895.

The Radziwills' *louche* lifestyle, however, was more difficult to handle. Stories of the broad-minded arrangement they had come to in their marriage—such as Lee's romantic dalliances with Onassis in his suite at Claridges, while Stas, who was very fond of hookers, entertained his latest *fille de joie* in the bar downstairs[10]—titillated smart dinner parties in London and New York.

It would not have been so bad if Lee had shown better judgment in her choice of lover; but to have embarked on an affair with "the Greek"—a man who courted publicity, and who was still engaged in a scandalous affair with the married Maria Callas; a man who had been indicted by the U.S. Justice Department for criminal conspiracy to defraud the United States of millions of dollars—was simply unconscionable, and Bobby regarded it as a betrayal of the whole family.

But if Lee's judgment was poor, her timing was even worse. Her decision to accept her lover's invitation to join him and Callas on a cruise aboard his yacht in the spring of 1963 was an unwelcome reminder to the Kennedys of how close Onassis had gotten to the family—and how much he might know about their affairs.

Plain reason had convinced the Radziwills' friends that their marriage was over, and that Lee had just as firmly set her sights on Onassis for husband number three as Onassis had selected her to be his new trophy mistress. But who was leaking the stories of their romance to the press?

Bobby blamed Onassis, of course. But another source might have been the "sexually cynical"[11] Stas Radziwill himself. Rumored to have financed his first property deals with money embezzled from the Polish Red Cross,* Radziwill was a man of considerable charm but little probity. Since he made little attempt to disguise his hurry to get out of his

---

* The story of Radziwill's embezzlement was well known among his friends (and enemies) in London. Gore Vidal claims that Radziwill "walked off with the entire capital of the Polish Red Cross, leaving an ancient Polish general to take the blame. Stas made money with the money; then, as prison doors began, creakily, to swing open, he restored what he had stolen, without interest." (Gore Vidal, *Palimpsest: A Memoir,* p. 372. London: Andre Deutsch, 1995.)

marriage to Lee, in order to devote himself to the pursuit of Charlotte Ford, the daughter of the auto magnate Henry Ford II, his friends suspected that he himself had provided some of the newspaper stories about Lee's capitulation to Onassis's wiles.

Nevertheless, at least one story, in Drew Pearson's *Washington Post* column, can be traced directly to Onassis—and it was the most disturbing from Bobby's point of view: *"[Onassis's] romance with opera star Maria Callas is reported on the rocks and his ambition is to be a brother-in-law of President Kennedy . . ."* [12]

David Karr, who played a shadowy role in Onassis's life, claimed that Onassis had asked him to plant the item with Pearson, for whom Karr had once been a legman, and still remained a close friend. "Ari wanted to light a little bit of a fire under the attorney general, and to show that he could reach into the very heart of Washington to do it," said Karr. [13]

In June, Bobby found the perfect excuse to get Lee off Onassis's yacht, which was cruising off the Italian coast, south of Rome. It was perfect because he knew that her departure would publicly humiliate Onassis, who was milking the publicity as the host of the First Lady's sister.

It was also an offer that Bobby knew Lee would not be able to resist; he asked her to fly to Germany to stand in for Jackie—now seven months pregnant and unable to travel—on the President's European tour. Jack, who had discovered the advantage of having an attractive woman at his side on the big occasion, was always pleased for an excuse to entertain his sister-in-law. According to Robin Douglas-Home, who had been one of Lee's intimate London friends during her marriage to Michael Canfield, she and Jack reveled in their shared promiscuity and sense of risk. They had, claimed Douglas-Home, been lovers almost from the moment they became in-laws; Canfield seemed to be in no doubt of it. "There were times when I think she went perhaps too far, you know? Like going to bed with Jack in the room next to mine in the south of France and then boasting about it," Canfield would later tell his old friend Gore Vidal. [14]

Although Onassis was convinced that Lee would sleep with her brother-in-law as willingly as she had slept with him in the south of

France, it was not a jealous anger he felt, for he had no sexual jealousy at all, but an anger that came from "the very heart of his self-esteem; that she had left his bed to sleep with one Kennedy at the behest of another made him very angry, indeed," according to Georgakis.

Lee accompanied Jack to his dinner for the West German president in Bonn, continued with him to West Berlin, and stayed by his side for the rest of the tour, including his sentimental journey to Ireland and an unplanned trip to England, where he visited his sister Kathleen's grave.*

The trip had been a personal triumph for Lee; cheering mobs had greeted Kennedy wherever he went, and Lee had been by his side through all of it. "She had a remarkable presence . . . At the same time, she was just one of the gang," admired one White House aide.[15] Few realized what sweet revenge it was for the woman who had lived in her sister's shadow for so long. "Why would anyone care what I do when there are so many more interesting people in the world?", she had asked a reporter not long after Jackie became First Lady. "I haven't done anything at all." Only her friend Truman Capote sensed the maladjustment that lay behind that *cri de coeur*. "My God, how jealous she is of Jackie: I never knew," he wrote in his journal.[16]

It would be John Kennedy's last European trip, his last public triumph. Until then, Lee had been on the periphery of events. But standing next to the president when he made his historic "*Ich bin ein Berliner*" speech in the shadow of the Berlin Wall, she would later say, was the most thrilling experience of her life.

Flying back to Washington on Air Force One, Jack asked her to do him one small favor: *Would she remarry Stas in a Catholic church?*

Both Jack and Lee knew that it was not at all a small favor he asked; it was a situation the Radziwills had been avoiding since the Vatican had by the grace of God—and the gall of Bobby—annulled her marriage to Michael Canfield.

---

* Kathleen was killed in a plane crash in France in 1948.

Despite some misgivings about his reliability, Jack Kennedy was fond of Radziwill. Stas had worked hard to win support for him in Polish-American communities in the Midwest in the 1960 presidential election, for little reward in the end.* Nevertheless, the point of the Catholic ceremony had nothing to do with the Kennedy's redemptory fervor. It was about politics, pure and simple: The Kennedy brothers wanted to make sure that Lee would not be free to marry Onassis before the U.S. elections the following November. "How could Lee refuse?" Stas would later wryly explain their predicament to friends. "Jack was her lover, her brother-in-law, and her fucking president!" [17]

No matter that Lee and Stas were no longer in love with each other, and were involved with other people; they both knew where the power rested. And one week after Lee returned from Berlin, the Radziwills dutifully and miserably went through a Catholic marriage ceremony in London.

"The Kennedys could accept me as Lee's lover: that was personal. What they couldn't accept was the idea that I might actually marry her: that was politics," Onassis told friends. According to one of Stas's London friends, Stas's agreement to renew their Catholic vows when all he wanted to do was dump Lee was "the most extreme gesture in a long career of sycophantic obedience to the Kennedys."

But why did he go along with an arrangement that was clearly against his own instincts and interests? "Bobby paid him," Onassis insisted. "Bobby dug deep to get Stas to go down the aisle with Lee a second time. Poor Stas would do anything for money." [18] Whether Bobby financed the Catholic ceremony or not, Robin Douglas-Home suspected that it was "the most plausible explanation on offer" for Stas's remarkable U-turn. "For people like Ari and the Kennedys, the answer to everything is

---

* Kennedy had planned to give Radziwill a role in his administration, but, according to Gore Vidal (*Palimpsest*, p. 372), when the FBI report was presented to him, reputedly weighing many pounds, and revealing Stas's careless management of the Red Cross funds in London, "Jack was plaintive: 'Should I read all this?' 'No, Sir,' he was told. 'Just don't make the appointment.' He did not."

money. If you have enough of it, you can buy anything you want," he told me.

If the Kennedys believed that the Catholic ceremony would impress upon Lee the need to be more temperate in her private affairs, they could not have been more wrong. A few weeks later, she was back on "the Greek's" yacht—without her newly rewedded husband.

"Lee went through the sham of that marriage service in London to please the Kennedys, but she wasn't going to sacrifice her whole future for the family," says one of her oldest friends in London.[19]

Whereas Onassis was physiologically sixty-three, sexually he was about forty, Lee told friends who discreetly inquired about such matters. Moreover, he knew "how to surround a woman with attention, and take note of her slightest whim," she said with undisguised satisfaction.[20]

Forty-eight hours after rejoining her lover aboard his yacht, Lee received the news that Jackie's baby, Patrick, had died three days after his premature birth at Boston Children's Hospital.

What duty could not convince Lee of, tragedy could. And whether driven by protocol, political imperative, or genuine sisterly sympathy, she hurried back to comfort her sister.

EIGHT

HOW COULD LEE REFUSE?

Jeanne, you can't make the beds and not
know the things that go on in them! Is
Mr. Onassis sleeping with Princess Radziwill?

—MARIA CALLAS TO A MAID ABOARD THE *CHRISTINA*

While the Kennedys mourned the loss of one child, Maria Callas was secretly celebrating the news that she was pregnant. Almost forty years old, she knew that it was probably her last chance of having Onassis's child, and her last hope of persuading him to marry her. For weeks she had kept her condition a secret, hiding her morning sickness from all but the devoted and wise Jeanne Herzog, her maid aboard the *Christina*.

Meneghini had denied her a child because he believed that motherhood would destroy her career, as well as his meal ticket. Nor was another child a priority in Onassis's life; it was not even an option, he had told her cruelly when she raised the possibility earlier, and she knew she would have to choose her moment to break the news to him with care.

Nevertheless, she also had cause for optimism. A few days earlier, "snooping around where I had no right to be," she had found a Cartier diamond bracelet with a note written in Onassis's hand, addressed simply to "my dearest, my sweetest love."[1]

Believing it to be a surprise gift for herself, she replaced it in its hiding place and prepared to be duly surprised when he gave it to her.

That, she decided, would be the moment to tell him about the baby.

Lee returned to Greece immediately after Patrick's funeral in Boston. "I was astonished Lee had not stayed with her sister. She repeatedly told us how terribly upset Jackie was by the death of her baby," Callas told her friend Nadia Stancioff.[2] But, as Onassis liked to say, and Callas was about to find out, the *Christina* is a ship full of surprises.

It would have been remarkable if by this time Callas had not at least suspected that Lee and Onassis were not more than just good friends, especially after Lee had returned so lickety-quick from her London nuptials, minus her groom. But it was only when Lee appeared wearing the bracelet Callas had set so much store by that she saw the light. Even then, she was still unable to accept that it was all over between her and Onassis.

"She said to me, 'Jeanne, you can't make the beds and not know the things that go on in them! Is Mr. Onassis sleeping with Princess Radziwill?'" recalls Herzog, who had been the *Christina's* senior maid for twelve years, and one of the longest-serving members of Onassis's personal staff. "I didn't want to lie to her, so I said that no one else slept in the big bed with Mr. Onassis, only Madame used that bed with him. Of course, I knew that he slept with the Princess elsewhere. It was a big yacht."[3]

Onassis loved to be seen entertaining Lee and Callas together—"he knew how their menage made tongues wag, and how much that upset Bobby and the Kennedys in Washington," said Yannis Georgakis. The evening Lee returned from the funeral in Boston, Onassis, Lee, and Callas dined together in Athens.

People in the know marveled at the sophistication and restraint of the two women flanking their lover. Neither woman resorted to hurt and angry looks; Lee acted friendly to Maria, and vice versa. Maria even managed to ignore the diamond bracelet on Lee's wrist that proclaimed Onassis's betrayal—and which Lee no doubt believed attested to the seriousness of his intent toward her.

Neither woman saw the sea of troubles into which they were sailing when Onassis made his fateful offer to put his yacht at the grieving First Lady's disposal. "Call her now," he insisted, after Lee had told them about Jackie's grief. "Tell her the *Christina* is hers for however long she wants it," he said, with the generous flourish learned from a lifetime of grabbing opportunities.[4]

Somewhat bizarrely, given the shape the Radziwill marriage was in, Lee told her sister: "Tell Jack that Stas and I will chaperon you. It will be perfectly proper . . . Oh, Jacks," she reverted to her sister's childhood nickname, "it will be such fun. You can't imagine how terrific Ari's yacht is, and he says we can go anywhere you want. It will do you so much good to get away for a while."[5]

Callas should have been alerted by the ardor of Onassis's invitation; Lee should have glimpsed a warning light in the alacrity of Jackie's acceptance.

But these danger signals were missed.

Jackie, of course, had already been aboard the *Christina,* when she and Jack were invited to meet Sir Winston Churchill in the south of France in the 1950s—when Onassis failed to ask them to stay for dinner with the Great Man.

The president was furious when he heard that Jackie had accepted Onassis's invitation. His secretary Evelyn Lincoln described him as "looking like thunder" the day he heard what Jackie had in mind for her convalescence.[6] But by this time, JFK, whom his friend Ben Bradlee considered to be "the world's champion male chauvinist pig"—a husband who barely even tried to hide his adulterous compulsions from his wife—had already sacrificed whatever claims he might have had on her loyalty on the *hauteur* of her disdain. All he could do was to beg her not to go.

According to Lincoln, the die was already cast: "Jackie had made up her mind, and that was that."[7]

The reason for her intransigence was perhaps not so much her husband's disinclination to stop his womanizing, but his failure to understand how deeply his unfaithfulness humiliated her.

And this was payback time.

On September 12, 1963, the Kennedys celebrated their tenth wedding anniversary with a family dinner at Hammersmith Farm. As an anniversary gift Jack gave his wife the catalog of a New York art dealer, inviting her to choose from it anything she wanted.[8] According to the London art dealer, and Jackie's friend, Billy Keating, it was this extravagant gesture that revealed the trade-off that he believed was now at the heart of their marriage.

Nevertheless, if Jack's generosity was calculated to weaken Jackie's resolve about the cruise, it did not succeed. When Franklin D. Roosevelt, Jr., who was shortly to become more involved in the matter than he wished, asked Bobby what they proposed to do about Onassis's invitation, the attorney general answered grimly, "Sink the fucking yacht."[9]

To have the First Lady floating round the Greek islands with her sister, her sister's husband, and her sister's lover—a man who had been sued for fraud by the United States government, and who was openly conducting an adulterous affair with the most famous prima donna in the world— was not the kind of photo opportunity the Kennedys needed in the run-up to their reelection campaign.

But it was a far more sensitive situation for the Kennedys than even those White House insiders who knew about Onassis's invitation suspected. For the president and his attorney general knew that "the Greek" was onto their affair with the late Marilyn Monroe. His "you fuck your movie queen and I'll fuck my princess" jibe could have left them little room for doubt on that score. Nevertheless, Monroe had been dead for over a year, and not the faintest hint of the true nature of her friendship with the Kennedys had yet leaked out. Beginning to breathe more easily, perhaps, Bobby decided to take the Greek bull by the horns and propose a fresh start in their relations.

"I'd been expecting Bobby's call," Onassis always began the story that would enter his personal mythology. "I was in bed with Manuela* at my apartment in Paris. She had been very good to me; I was in a good mood when I took Bobby's call." (In some versions, Onassis claimed that Manuela was still performing fellatio on him throughout his conversation with Bobby.)

He said that Bobby was sweet reason itself. "He told me that he and the president would withdraw their objections to my affair with their sister-in-law—even to our marriage if that was what we had in mind—if I would call off the cruise with the First Lady," he said. Bobby might have been the second most powerful man in the United States, but Onassis knew at once that he had already failed to persuade Jackie to change her mind about the cruise. Why else would he be offering terms to Ari? He went to the heart of Bobby's weakness: "I said, 'Bobby tell me: Does Mrs. Kennedy know about this offer? Does she have any say in the matter?' " [10]

Even diplomacy has its limits, and in Bobby they were usually reached sooner than later. He had played the only hand he had, and Onassis had brushed it aside with disdain. Bobby went insane with rage, according to Onassis. "He said he would destroy me," he later told Yannis Georgakis. "I said, 'My boy, you don't frighten me, I've been threatened by experts.' He said, 'What's in the past, you Greek sonofabitch, will be nothing compared to what's in store.' " [11]

But if Onassis himself was not intimidated by Bobby's rage, the threat that Onassis's tankers would be banned from the United States disturbed Costa Gratsos. Although not even the attorney general of the United States could flout international shipping agreements on a private whim, even when channeling the president's own fury, Gratsos certainly knew a man who could: Jimmy Hoffa. Ordinarily, of course, the prospect of the Teamsters' president joining forces with Bobby Kennedy to keep Onassis's tankers out of American ports would seem to be high on a list of

---

* Manuela was reputedly the most sexually gifted Madame Claude girl in Paris, and the one Onassis selected to initiate his son, Alexander.

things that could never happen. Nevertheless, stranger accommodations have been made when a mutual interest is at stake and powerful personalities are involved.

What troubled Gratsos, and rendered the unlikely alliance less remote, was the fact that Onassis had agreed to invest ten million dollars in Spyros Skouras's new fleet of container ships, which would cut labor costs by up to sixty percent, and jeopardize thousands of longshoremen's jobs. It was a dangerous time on the waterfront; labor racketeers were looking for opportunities to wreck honest union deals and strengthen their own power bases. Threats to Skouras's life were being taken seriously in Washington; secretary of commerce John T. Connor had offered him around-the-clock protection.[12]

Whether or not Bobby really would have thought of using Hoffa to block Onassis's tankers because of his investment in the Skouras fleet is a moot point, but the idea had a symmetry of retribution that seemed obvious to Gratsos's Machiavellian mind. Onassis, too, saw the risk, and arranged a meeting with Hoffa at the Pierre Hotel in New York to tell him personally that he was pulling out of the Skouras deal. In gratitude Hoffa offered Onassis some advice on how to deal with Bobby Kennedy. "Wink at him. The kid's a fucking faggot," he said. "He dressed like a girl until he was ten years old!"[13]

Shortly after his meeting with Hoffa, Onassis had lunch with Rupert Allan in Los Angeles. A Rhodes Scholar, a successful newspaperman at the *St. Louis Post-Dispatch* and *Look* magazine before he became a Hollywood press agent in the fifties, and eventually Monroe's personal publicist, Allan was a polite, fastidious man who handled the press with the same aplomb he used to deal with impossible movie stars.

He had just returned from Monaco, where he had been visiting his old friend and client, Grace Kelly. Onassis's break with Monaco had not been exactly friendly; there was no love lost between Onassis and Kelly, and Allan imagined that Ari wanted to pump him about what was going on behind palace doors. He was surprised when Onassis wanted to talk not about Kelly but about Monroe. "He asked me if I believed Marilyn had really killed herself. I told him that both her nature and mine had given rise to thoughts of suicide in the past, and that we had a pact to call each other if ever we got into that frame of mind again." The fact that Monroe had not called him the night she died inclined him to doubt the suicide theory.* "Ari said, 'Yes, that's what I think, too, Rupert. She didn't die by her own hand.' "[14]

He then asked whether Allan knew about a bugging operation begun against Bobby Kennedy and Monroe at her home in Brentwood a few weeks before she died. "I knew that Marilyn had exchanged body fluids with both Bobby and Jack, but that was the first I'd heard that Fifth Helena Drive might have been bugged," Allan later recalled. Onassis asked whether he could get hold of the incriminating tapes for him. Although he was a man who viewed life through the prism of Hollywood mores, Allan declined to get involved; that kind of work was out of his line. Anyway, he added, if they existed, they would cost a great deal of money. Onassis laughed. "He said, 'Rupert, Rupert,' " Allan recalled his reply with relish, " 'my dear fellow, hasn't anyone told you? I *have* a great deal of money.' "†

Allan had known Onassis since 1956—when, as an MGM publicist, he had accompanied Kelly to Monaco to handle the press at her marriage to Rainier; he knew about Ari's obsession with Bobby Kennedy's private life, which Allan believed had acquired the attributes of a personal vendetta. Although he knew that Monroe had been seeing Bobby Kennedy up to the day she died, Allan was cynical about Onassis's story of tapes, which he dismissed as "Ari's wishful thinking." He simply couldn't believe that the attorney general of the United States could be "stupid enough to let himself be caught on tape in his girlfriend's bedroom."

---

* Monroe had, in fact, sent a message asking Allan to call her a few days before she died, but he had just gotten back from Monaco and had jet lag as well as a bad case of bronchitis, and did not get back to her. He knew, he would later tell friends, that if he spoke one word to her, "she would insist on coming over with chicken soup and aspirin. And I was really too sick for that."

† It was, in fact, one of Onassis's favorite lines.

Later, when it appeared that Bobby *had* been that stupid, Allan believed that his problem was hubris, an arrogance derived from money, success, and power: "Bobby Kennedy was just like Onassis in that respect. He didn't believe that the ordinary rules applied to him."

Nevertheless, Allan had remained skeptical about the existence of the tapes until 1985, when the British investigative writer Anthony Summers, in his ground-breaking biography of Monroe,[15] revealed that a wiretap expert name Bernard Spindel had bugged Monroe's home—as well as Peter Lawford's house, where some of Monroe's trysts with Bobby had also taken place—in the weeks before her death.

Unless Allan's recollections are inaccurate, or his conversation with Onassis had been much later than he believed, which seems unlikely, Onassis clearly knew about the Monroe–Kennedy tapes long before Tony Summers revealed their existence. Accounts of whom Spindel was working for when he bugged Monroe remain contradictory and inconclusive, but one of his main patrons was Jimmy Hoffa—the man Onassis had visited shortly before his meeting with Rupert Allan.

What had Onassis intended to do with the tapes if he had found them? According to Allan's account, Onassis told him that when he was a young man in Smyrna, he had asked a Turkish captain why powerful nations like Britain, Germany, and the United States had stood by and let the massacre of the Smyrnan Greeks and Christians continue? "Because they can't afford to upset us," the Turk told him; his country was only a little nation, but it now controlled a major strategic crossroad in Asia Minor.

"Onassis said that he realized then that all you need is one golden apple—a single apple that somebody else wants—and you have control. I said, 'Ari, that's blackmail.' He said, 'No, Rupert. That's business.' I still don't know whether he ever got hold of the tapes—his golden apple—but if he had, he would have controlled Bobby Kennedy, that's for sure," Allan told me in 1991.

NINE

A CHARMING PSYCHOPATH

**All Greek men beat their women.
He who beats well, loves well.**

—ARISTOTLE ONASSIS

F ranklin Delano Roosevelt, Jr., was a popular figure in the White House, but he had not achieved the recognition he felt he deserved in the administration.[1] A junior government post as Undersecretary of Commerce, and a White House dinner invite for his friend, Italian car magnate Gianni Agnelli,* seemed modest rewards for contributing so much to John Kennedy's critical primary victory in West Virginia, where the Roosevelt name still had magic.† His hopes must

---

* Roosevelt was Agnelli's Fiat distributor for the East Coast.

† FDR Jr. campaigned through West Virginia with JFK, linking the Kennedy and Roosevelt names in a state where his father's New Deal was still holy writ. West Virginia journalist Charles Peters says he was "almost God's son coming down and saying it was all right to vote for this Catholic, it was permissible, it wasn't something terrible to do. FDR Junior made it possible for many people to vote for Kennedy that couldn't have conceived of it as a possibility before" (Oral History, JFK Library). When the campaign got tough, Bobby persuaded Roosevelt to release to the press material smearing Jack's rival, Hubert Humphrey, who had been medically rejected for military service in World War II, as "a draft dodger."

have soared, however, when the president summoned him to the Oval Office in September 1963.

It was Bobby Kennedy, in fact, who told FDR Jr. what the president had in mind. "We're sorry you didn't get the navy job, Frank, but we can fix you a great two-week cruise in the Aegean."[2] It was a cruel jest. Roosevelt had wanted to be Secretary of the Navy, but the appointment had been quashed by Robert McNamara when he became Secretary of Defense and demanded the right to pick his own men. ("I hear you're going to name Frank Roosevelt Secretary of the Navy?", a newsman had confronted McNamara. "The hell I am," McNamara replied, and that was the end of Roosevelt's dream.[3])

Roosevelt's presence on the cruise, along with his wife Suzanne, would add some respectability, the president told him, having explained the situation.

Nevertheless, Roosevelt would have been a curious choice if respectability is considered a requisite in a chaperone. He had been renowned at Harvard for smashing cars and staging wild parties. After his first marriage to heiress Ethel Dupont, they lived with his parents at the White House, where, like Kennedy after him, he entertained his lovers (and drove poor Ethel crazy). He had even propositioned his brother Elliott's wife, Patty, simply because, as he told her with alarming frankness, "you are the only sister-in-law I haven't had."[4]

Nevertheless, Americans trusted the famous name. On September 26, the evening before Roosevelt left to join the *Christina* in Athens, he was briefed by Bobby. To throw the press off the scent, he was to travel via Egypt, where he would meet with President Nasser on some "very hush-hush matters that President Kennedy wanted me to discuss."[5] As well as pursuing the chimera of respectability, the attorney general told him, Roosevelt must keep "the Greek" as far away from the First Lady as possible, especially if there were cameras about. He did not want to see Jackie and that Greek sonofabitch grinning up at him when he opened the *Washington Post* at breakfast one morning.

Roosevelt had known Onassis for more than twenty years, and did not need Bobby to tell him how ill advised it was to permit such a man to play host to the First Lady. "I thought he was a charming psychopath the first time I met him, and I never changed my view. People who trusted him usually got hurt," FDR Jr. would later tell a friend.[6]

A vacation that had been presented as one of the perquisites of middling office had become something far more tricky. Roosevelt began to hedge. "Although I was a very old and personal friend of both the President and Mrs. Kennedy," he later told author David Heymann, "I didn't feel it was the right place for a government official to be, away from his public duties in Washington, floating around in the Eastern Mediterranean." At this point, he was fifty-three years old; he still had political aspirations. "I was concerned that this might be called sort of a playboy cruise, even with my wife there, and I didn't need anything like that."[7]

But Bobby knew that men who dance attendance were always willing to learn new steps, and the following morning Franklin D. Roosevelt, Jr., accompanied by Suzanne, dutifully albeit reluctantly left Washington on his journey to Athens, and, as it turned out, political oblivion.*

That Onassis had become a man whom even the president of the United States could not afford to ignore proved how far he had come in the twenty years since he first arrived in America. Unaware of how much his stories exposed the web of lies, deceit, and manipulation that were at the heart of his success, it was a journey Onassis liked to relive for his friends.

By the start of World War II, he had parlayed his Argentine tobacco profits into a fleet of freighters and three of the world's largest tankers, and the war offered fat profits for neutral shipowners prepared to run the risk of hauling Allied supplies across the North Atlantic. But for Onassis it was also a battle to keep his biggest and most valuable tankers from being commandeered or impounded by one side or another.†

---

* When John F. Kennedy was assassinated a few months later, Roosevelt was left out in the cold. He ran for mayor of New York and then for governor on the Liberal ticket in 1966, but with ever diminishing votes, and died in 1988 on his seventy-eighth birthday.

†To emphasize their neutrality, Sweden had impounded for the duration all foreign-owned ships sailing under its flag, or built in Swedish yards, including Onassis's

The strain was immense. "A blood vessel or something has broken in my throat," he wrote from London in 1940 to his Norwegian mistress Ingeborg Dedichen in Paris. "I can't talk without the taste of blood in my mouth, and the smell of it in my nostrils. And all the time I am facing the possibility of not keeping a dime, of being left penniless because of these adventurers called politicians."[8]

Ingeborg, whom Onassis had met aboard a ship taking them from Buenos Aires to Genoa in 1934, wasn't just along for the ride. The daughter of Ingevald Martin Bryde, one of the most respected shipowners in Norway, she had brains as well as a Garboesque beauty.* She relished her role as Onassis's social mentor, as well as his mistress, and although her friends were appalled by what one politely called his "Greek vivacity"—summoning waiters with clicking fingers, or hissing like a snake—Ingeborg knew that it was his indifference to what people thought of him that gave him his energy and his style.

She loved the sight, sound, and touch of him, too: the way his "heavy head emerged directly out of the muscular shoulders of a docker . . . there was nothing of the thoroughbred about him: he looked like a simple stevedore out of Asia Minor."[9] Yet his skin had "an odor, a texture, a warmth and velvet softness which was beyond compare, of which I never tired," she would recall. Before making love, he "liked to lick me between the toes, carefully, like a cat . . . he would embrace every part of my body and cover me with kisses before devoting himself to the feet he adored."[10] Onassis recalled it less lyrically: "We fucked a lot. We fucked all the time."[11]

Ariston—which had been the world's first 15,000-ton tanker—and his 17,500-ton *Buenos Aires,* still in her fitting-out berth in Goteborg, and registered under the pro-Axis Argentine flag. The Norwegians had also commandeered his *Aristophanes.*

* Dedichen had persuaded influential friends in Sweden and Norway to vouch for Onassis when he needed the Goteborg shipyard that was building the *Ariston* to drop the punitive "Greek clause" requiring at least 50 percent cash and a maximum five-year credit line. It was also her influence that enabled him to pay the first 25 percent in three stages during construction; the remaining $600,000 would be spread over ten years at 4.5 percent interest. "No Greek had ever got a deal as good as that, and he had Mamita [meaning Little Mother, Onassis's pet name for her; she called him Mamico] to thank for it," Costa Gratsos told me.

As the war clouds darkened in Europe and he fought to recover his impounded tankers and avoid future seizures, the state of his mind can be judged by his letters to Ingeborg. "I know I lose my temper," he wrote to her in Paris from the Savoy Hotel in London. "I know I get mad with you. But I do love you . . . People talk about my strength of caracter [sic], my volonté in business. But I have been a coward where we have been concerned. I always told myself: tomorrow, next month, next year . . . and here we are, after six years, another separation, another heart-break. I've been so blind." [12]

But financially and physically he was on a knife edge, and his tone moved from tenderness and concern to exasperation and self-pity. "Don't you understand that I'm now having to contemplate the possibility of being left with almost nothing!" he scolded her in another letter. Most of the Greek shipowners had quit London for New York, where he complained they were "living on velvet in safety, even the very stupid ones are making fortunes." [13] His exhaustion and his bitterness showed in letter after letter. "All my work of nearly twenty years, all the sacrifices and the abominable and abnormal living of all those years, could turn into a great nothing! Most men in my situation would now give up, Mamita, they would be perfectly understood if they committed suicide!" [14]

Instead, on July 1, 1940—the day U.S. ambassador Joe Kennedy told former British prime minister Neville Chamberlain that his country would probably be "beaten before the end of the month" [15]—Onassis quit London for New York aboard the Cunard Lines' SS *Samaria.*

He made no secret of his fear as the liner zigzagged across the Atlantic to avoid marauding U-boats. Clutching his briefcase, which contained contracts, deeds to his ships, proof of everything he owned, he slept on the deck closest to the lifeboats. "During ten days and nights I have been with the same clothes on and I slept on the divan of the smoking room first to be ready if anything happened . . ." he wrote Ingeborg when he reached New York. [16]

His first act in New York was to dispatch his cousin and brother-in-law Nicos Konialidis—who had taken the rap for Onassis's insurance fraud in Genoa [17]—to Rio de Janeiro to recover the *Aristophanes,* which had been commandeered by the Norwegians. Unless the Norwegians agreed to pay him a million dollars, the tanker, a precious Allied asset, would spend the rest of the war in Brazil tied up in lawsuits.

"They did all they could not to pay me, but as I had a mortgage on the ship and having arrested her in Rio through Nico they had finally to pay me . . . so while all the other owners have to submit themselves to the (Allied shipping) mission's wishes, thanks to my mesures [sic], I found myself in a very privileged and unic [sic] position which saved me," he wrote.[18]

On September 4, he sent Ingeborg an account of his wealth: "Speaking in figures out of 8 millions [sic] dollars,* I have managed to save two and a half for which I am very thankful and satisfied . . ."[19]

In New York, Onassis realized how much he missed—and, more importantly, still needed—Ingeborg. But now that they had been separated by war, she was determined to make the break permanent. She knew he would protest, for she was taking something away from him, and it was not in his nature to lose.

He turned on her all the power of persuasion that had carried him from the ashes of Smyrna to a suite on the thirty-seventh floor of the Ritz Towers in Manhattan. But neither he nor she was a good letter-writer, and the correspondence—in English, the language that had both united and divided them from the beginning—simply exacerbated the complications and misunderstandings which had been the pattern of their affair from the start.

"It is funny Mamita but my whole mentality and caracter [sic] has changed immensely," he assured her. "I like myself much better and I am certain though quiter [sic] I must be now a far more pleasant person to be with than before . . . To this I must say your attitude has been the greatest factor . . . In the past I would have felt too proud to confess such a situation but now I can see life quite differently . . . on the contrary I like to confess and admit my defeat in this chapter of my life . . . Just think Mamita of my life since the age when I was six lost my mother

---

* More than $100 million in today's money.

thereafter a stepmother with one of the most difficult fathers during nine years until I became fifteen then all alone found myself in S. America without the warmth or the affection of anybody . . . I am thirty-four and I am still living like a vagabond. Just think of all these last five to six years living continuously on trains boats flying all alone in hotel rooms not knowing how to spend endless evenings not having a person to tell me a single comforting word . . ."

His letters made her weep. But even when they came from his heart she knew they were full of the lies he always told about himself.* He did not understand her at all. He did not know why she laughed or wept. He examined himself, exposed his own guilt, confessed his failures, but he had no idea about her needs. No matter what he promised, she knew that he would always want the lion's share . . . even of her sadness.

Nor had she forgotten how quickly his violent temper could erupt. She admitted her fear to him now that there was an ocean between them: "The idea of not to have to be affraid [sic] of you became so intense, all-most [sic] wild, that I wanted nothing else. Please try in friendship to understand me, please give me your real friendship, please try not to hurt me . . . even by vengeance you would like to in the moment . . ." she pleaded with him.[20]

He replied that she should also remember the good times. He had contacts in Washington who would fix the necessary visas to get her out of France. "My plan in asking you to come here is 1) to marry 2) to settle down at any place you like, promissing [sic] to have you a home ready not to stay in hotels untill [sic] the situation in Europe is settled. I don't tell you anything about love and happiness which I am longing to give because that should go without saying."[21]

---

* He had been twelve when his mother died; twenty-two when he went to Argentina; he was forty years old, the same age as she was, not thirty-four. Applying for a reentry visa to the United States, he stated that he was born in Salonika, Greece, on September 21, 1900. He gave his nationality as Argentine and was in possession of a valid Argentine passport, number 701014, issued by the Buenos Aires police department and valid for two years. He gave his legal address as Reconquiesta 336, Buenos Aires, and his address in the States as the Ritz Towers, 57 Park Avenue, New York City.

She capitulated, as she always had. Pitifully, she set out her terms:

"Wherever we may live—New York, Athens, France—I want a place to live in, big enough to be independent, so that I can receive my friends, people you may like or dislike, without needs to bother about you. If such a place difficult to find, to live in a flat alone."[22] The apartment, she went on, must be 100 percent hers. She would also expect to receive a regular income, for which she would not have to present accounts to him, be provided with a good servant, be free to go to "conserts [sic] and teatres [sic] I like, if you like to come with me I shall prefer, but no troubles in the last moments, or changing of program!"

He cabled back:

Accept all your conditions. Amour tendresses Onassis

The affair followed the pattern of any relationship conceived by obsession on one side and capitulation on the other. They made a striking couple: tall, slim, her blonde hair styled in waves like Carole Lombard's, Ingeborg (or Inge, as she was called by her new friends), was the perfect partner for the small, sleek Onassis, who had—according to Bill Carter, an early New York companion—"an air of controlled violence" about him, which "some women found sexy."[23]

He had made an interesting circle of acquaintances in New York. Celebrities, socialites, movie people—Otto Preminger; Spyros Skouras; Gloria Swanson; a lively blonde actress named Constance Keane; the President's son, Franklin Roosevelt, Jr.—mingled with brokers and bankers at his parties. Their cottage at Center Island on the north shore of Long Island became a popular weekend meeting place. Inge was a fine hostess, and even the establishment Greeks who had cold-shouldered him in London—including the formidable Livanos and Embiricos families—began dropping by for drinks.

But shortly after Pearl Harbor, Onassis told Inge that he was seeing other women. He didn't want to end their relationship, but he wanted to play the field. One reason for his change of mind was sugar heiress Geraldine Spreckles.

Although he was nearly twice her age, Onassis fascinated Spreckles as much as she excited him. His sharp mind, his *joie de vivre,* the stories he told, his sheer energy and ego bowled her over. They became engaged to be married. "He was simply outrageous. Not being found out was the same to him as telling the truth," she told me.[24]

Spreckles was offered a Warner Brothers' contract and moved to California; Onassis took a suite at the Beverly Hills Hotel. His engagement to the young heiress did not cramp his style, however. He dated Paulette Goddard and the French sex star Simone Simon (from whom he reckoned he contracted a venereal disease); he caught up with Constance Keane, who had become Veronica Lake, and embarked on a discreet short-lived affair with Joe Kennedy's former mistress, Gloria Swanson.*

But when Spreckles stood him up at the altar, Onassis returned to Ingse in New York, where she had taken an apartment two floors below his in the Ritz Towers. His drinking had become excessive, and the slaps that he had occasionally given her in the past became harder and more frequent. He admitted to taking a sexual pleasure in the violence. All Greek men beat their women; he who beats well, loves well, he said.

Once he beat her so badly that one of her eyes was pulped to "black butter."[25] The following day, when he saw how badly he had hurt her, Onassis drove her to their cottage, dismissed the servants, and nursed her until the injuries healed.

Ingse feared that one day he would go too far and kill her—or, if not her, somebody else. Her concern for the man she had lived with and loved for twelve years was as palpable as her own despair. But she knew that it was over between them—by this time Onassis had met Tina Livanos—and in 1946, shortly before Onassis married Tina, Ingse quietly returned to Europe.

---

* Shortly after Onassis began his affair with Gloria Swanson, she was offered the lead in a movie called *Father Takes a Wife.* "The story was a light comedy about a famous actress who marries a shipping tycoon," she observed wryly in her memoirs, but omitted to mention her affair with Onassis. (*Swanson on Swanson,* p. 466. New York: Random House, 1980.)

TEN

THE HUNGRY LITTLE GREEK

**The hungry little Greek
can do anything.**

—JUVENAL, c.e. 55–c.e. 140

J ackie did not contact Onassis directly to accept his offer, but on September 18, 1963, the White House announced that the First Lady would spend the first two weeks of October in Greece, "vacationing and convalescing."[1] There was no mention of Onassis, no mention of his yacht, no hint of any cruise. In New York, Costa Gratsos, who was as vehemently against the cruise for his own reasons as Bobby was for his, read the story in the *Times* and began to worry.

The following morning, Gratsos flew to Athens and drove straight from the airport to Onassis's villa in Glyfada, a coastal suburb ten miles southwest of the city. Situated directly beneath the flight path of Athens international airport, it was the oldest and ugliest of the nine or so residences Onassis owned around the world. But it was the home Onassis loved best; he had built an identical house beside it so that he could have his sister Artemis and her husband Theodore, a physician, close at hand.

Gratsos understood Onassis better than anyone alive. The same age as "the hungry little Greek," as he affectionately called Ari, Gratsos

belonged to one of the oldest Greek shipping aristocracies; his mother
was a Dracoulis, one of the most respected families from the island of
Ithaca. His weathered face and burly physique concealed an intellect that
had been honed at the University of Athens and the London School of
Economics. He never sought the spotlight, or yearned, like Onassis, to
see his face on the cover of *Time*. It was enough that his peers knew who
he was and respected the part he had played in Onassis's success. They
had first met—"at a night club," remembered Gratsos; "in a brothel,"
insisted Onassis—in Buenos Aires in the 1920s. "Costa knows every
crime I've ever committed," Onassis liked to say.

There were no other guests for dinner on that September evening. It
was always agreeable for them to sit in each other's company, and the
evening would have followed the rules of a familiar ritual: Removing their
jackets, they would eat in Onassis's study, drink whisky rather than wine;
the food, in contrast with the elaborate French cuisine Onassis served his
guests on the *Christina,* would be simple—probably taramasalata and
grilled fish cooked by the housekeeper. (Onassis no longer loved to eat as
he did when he was young and when good food was cheap and plentiful
in Buenos Aires and Paris; now he preferred plain, digestible food.) Gratsos and Onassis would have talked with the unguarded ease of men in
their fifth decade of friendship. Even so, Gratsos knew that Onassis would
have to be handled with care: "Always respect his pride. He is not a
rational man when his pride is threatened," he would tell Johnny Meyer
the day Onassis hired Meyer away from Howard Hughes.[2]

And it was this tricky question of Onassis's pride that Gratsos had
flown from New York to talk about.

For the past nine months they had been putting together a deal with
François "Papa Doc" Duvalier, president of Haiti, to turn Port-au-
Prince, the capital of the impoverished former French sugar colony, into
a Caribbean Monte Carlo, or a new Havana. The most ambitious business transaction since his infamous Jiddah Agreement eight years earlier,
the deal would include an ultimate investment of over $375 million
(over two billion dollars today) in hotels, a new casino, a harbor develop-

ment scheme, an oil refinery, and the establishment of a flag of conve-nience state in which to register his tankers under even more favorable terms than he was getting in Panama.

They were also exploring the possibility of searching for oil and natu-ral gas reserves. It was a new venture for them, and they had teamed up with George de Mohrenschildt, also known as Jerzy Sergius von Mohrenschildt. Technically entitled to call himself a baron, he had met Onassis three years earlier in Panama at the home of Onassis's lawyer and friend, Dr. Roberto Arias.

De Mohrenschildt, who had already negotiated a geological survey contract with Papa Doc, was a tall, handsome fifty-one-year-old with an exotic past and, although nobody could possibly have known it then, a notorious future. He was in the Social Register, flew with the jet set, and claimed to have useful ties with the CIA, who would be able to help them in Haiti when the time was right, he told Onassis.* De Mohren-schildt had also known Jackie and Lee since they were young girls living in New York with their mother, Janet Lee Bouvier. He claimed that he and Janet had been lovers, which was not improbable given his penchant for other men's wives and the unhappy state of Janet's marriage to "Black Jack" at that time.†

Gratsos believed that de Mohrenschildt, in spite of his aristocratic air of self-confidence, was in way over his head in Haiti; there was some-

---

* According to Warren Commission Document No. 1012, dated June 3, 1964, and declassified May 31, 1977, Richard Helms, formerly CIA Deputy Director of Plans, admitted that in 1942 the Office of Strategic Services, forerunner of the CIA, consid-ered de Mohrenschildt for employment but did not hire him because of allegations that he was a Nazi agent. According to the Helms memo, the CIA first established contact with de Mohrenschildt in December 1957, after he returned from a mission in Yugoslavia for the International Cooperation Administration. The CIA had several meetings with de Mohrenschildt at that time and maintained "informal, occasional contact" with him until the autumn of 1961.

† Although de Mohrenschildt would also boast about his friendship with the Bouviers in his Warren Commission testimony, he claimed only that he and Janet were "very close friends. We saw each other every day."

thing about the smooth petroleum engineer that Gratsos could not pin down, and it nagged at him. Gratsos suspected that Onassis's enthusiasm for their putative Haiti partner had "more to do with what de Mohrenschildt knew about Jackie than what he knew about oil."[3]

Gratsos's concern was understandable. The Haiti project had been his idea, it was his baby, and he felt that Onassis, carried away by his affair with "Princess" Lee, was letting his social life get in the way of work. He wanted to renew his old partner's gung-ho enthusiasm for it.

The deal was fraught with enough difficulties and risks anyway. Not the least of these was the fact that the Kennedy administration had rejected the legality of Haiti's 1961 election, cut off U.S aid, and branded Papa Doc a murderous tyrant, which he clearly was. Moreover, shaken by how much Onassis's affair with Lee had offended President Kennedy, and even more importantly, given the nature of the man, by how deeply his invitation to Jackie had angered her brother-in-law, the U.S. attorney general, Gratsos knew that "one sure way to lose the Haiti deal was to win the battle with Bobby Kennedy."[4]

Over dinner, he spelled out his worries to Onassis. They had both taken a gamble on the Haiti venture and just when the risk seemed about to pay off, did it really matter whether Jackie went on the cruise or not?

"It matters," Onassis told him, "because Bobby has made it matter."[5]

ELEVEN

LACE HAS LANDED IN ITHACA

So this it seems is what it is to be a king.

—JACQUELINE KENNEDY,
AFTER ALEXANDER THE GREAT

I n the early hours of October 4, 1963, wearing dark glasses and a woollen coat of devotional simplicity (actually a costly creation of her White House couturier, Oleg Cassini), Jackie, accompanied by Lee and Stas Radziwill, rendezvoused with the Roosevelts in a side street behind Constitution Square in Athens. Two embassy limousines took them to Piraeus, the ancient port of Athens, where the *Christina* waited. "It was all very cloak and dagger," recalls Mrs. Erasmus Helm Kloman, the former Suzanne Roosevelt. "Jackie had a code name: Lace. It was marvelously clandestine."[1]

But Onassis felt humiliated by the furtive atmosphere that surrounded Jackie's arrival. His brother-in-law Theodore Garofalides, a once able physician who had become acquiescent in Onassis's service, exacerbated the situation by telling him that one of her Secret Service men had said that Bobby had ordered the security blackout to ensure that the press didn't get photographs of Onassis welcoming Jackie aboard the *Christina.*

We have only Onassis's account of their first words to each other that night, but they have a Stanley and Livingstone simplicity that seems to substantiate the truth of the exchange: "I welcomed her aboard. I told her I was pleased that she had been able to come. She said, 'I never intended it otherwise, Mr. Onassis.' "[2]

It was the moment he realized that Jacqueline Kennedy had a will as determined as his own, Onassis told me. Before he went to bed that night, he called Gratsos in Paris and told him triumphantly: "Lace has landed in Ithaca."* Gratsos felt as much dismay as any Kennedy to hear this. "I hope you're not going to regret it," he said.

"I'll make the waves, Costa," Onassis told him. "Let's see if Bobby Kennedy can walk on them."[3]

When Suzanne Roosevelt began her journal of the cruise with the words "There is absolutely no luxury which cannot be had as a matter of course," it was nothing but the truth.[4] Eight varieties of caviar, the finest vintage wines, cheeses, and exotic fruits had been flown in (courtesy of Olympic Airways) from all over the world; the crew of fifty-eight was complemented by two hairdressers from Paris, three chefs (French, Italian, Greek), a Swedish masseuse, and a steward and private maid for every stateroom. There was also a small orchestra to serenade them while they ate, and later to dance to.

Ex-king Farouk had called the *Christina* "the last word in opulence." But its mixture of luxury, vulgarity, and great beauty must have puzzled Jackie. How could a man with taste enough to use floorboards taken from a stately English home decorate the same room with models of ships that, at the push of a button, lit up and moved around like a carnival amusement? How could a man who had the finest collection of

* Of the nine staterooms aboard the *Christina,* each named after a legendary Greek island, Ithaca was the one reserved for his special guests. Previous occupants included Winston Churchill, Greta Garbo, and Maria Callas.

Greek classics she had ever seen in a private library also cover his bar stools with the skin of whale scrotums?

But, according to Onassis, Jackie was overwhelmed by the *Christina,* and told him: "So this it seems is what it is to be a king." It was what Alexander the Great had said when he entered the tent of Darius, the Persian king he had just defeated. It is unlikely to have been a spur-of-the-moment remark, and reveals how carefully she must have prepared herself for the visit, and how much she wanted to impress her host.*

The guest list had grown to nine by the time the *Christina* sailed out of Piraeus on the evening of October 4, headed for the straits of Istanbul. In addition to the First Lady, the Radziwills, and the Roosevelts, there were Princess Irene Galitzine, the Russian-born dress designer; Accardi Guerney, a friend of Lee's; and Artemis and Theodore Garofalides. "A hard mix," Suzanne Kloman would recall the names more than thirty years later.[5] "The chemistry was all wrong," admitted Theodore Garofalides.[6] But Stas Radziwill knew that it was more than a matter of shipboard mésalliance. "Everyone was caught up in some kind of deception," he told friends after he left the yacht in Istanbul.

Understandably, Stas had wanted to decline the invitation. He had told Onassis he couldn't go; he had told Lee he wouldn't go; but Bobby told him he had to go. Nevertheless, like a touch of insanity in a well-bred family, Stas's delicate situation was never mentioned. But in his old London friend David Metcalf's opinion, Stas's allegiance to his powerful in-laws and deference to his wife's generous lover were obligations he

---

* According to Charles Bartlett it was Jackie "who dug up a lot of the quotes that Jack started dropping in his speeches" in the fifties. "What she would do," said Kennedy staffer Meyer Feldman, "is make suggestions to Jack—ideas on positions he might take, poetry he might recite, historical references. If he spoke to a French-speaking group in New Hampshire, she came up with a phrase that she felt he could use" (Edward Klein, *All Too Human: The Love Story of Jack and Jackie Kennedy,* p. 226 New York: Pocket Books, 1996.) She also provided the same service for Bobby, finding him the quote he read after John Kennedy's assassination: "When he shall die / Take him and cut him out in little stars / And he will make the face of heaven so fine / That all the world will be in love with night / And pay no worship to the garish sun."

found harder and harder to sustain—especially now that he had set his sights on the Detroit heiress, Charlotte Ford.

Very soon, the group that had been hastily rounded up to give a look of respectability to the cruise began to bicker. Frank Roosevelt vented his hurt pride on Stas Radziwill, whose presence confirmed the politician's worst fears that it would be seen as a "playboy cruise"; the Polish prince vented *his* spleen on the Russian princess. "She hates me because my title is older than hers," he told Garofalides.[7] But Galitzine hated him not because of the antiquity of his rank, but because of "the despicable way he sucked up to his wife's lover."[8] Artemis did not approve of Lee: She believed that her brother needed a good Greek woman who would stay with him no matter how badly he behaved; *he needed Maria, not some fake American princess.*

For the first couple of days, Onassis barely left his stateroom. While his guests dined on *langouste thermidor,* he ate French bread and cheese in his study. He said that it was to remind himself that he had not always been rich. It was a romantic idea, but the truth was that his stomach could no longer handle rich food, especially when he was nervous or angry. It was Jackie who insisted that he spend more time with his guests, and that evening, he joined them for dinner and did not hurry off afterward.

He had lied, of course, when he said that Jackie would decide where they would go. This was *his* voyage, and they would follow the route he always took with those he wanted to impress. Brought up on the myths of ancient Greece, he saw himself as a latter-day Odysseus; the route through the islands—from Smyrna, the birthplace he shared with Homer, to Ithaca, where his hero had ended his travels—was a panorama of his life. Tina, who had made the journey many times—with Greta Garbo, with the Rainiers, with Winston Churchill—had called it his "ultimate ego trip."[9]

It had been on one of those trips that he had fallen in love with Maria Callas. Would history repeat itself? It was soon apparent to everyone on board that Jackie and Onassis had certainly hit it off tremendously.

Jackie was fascinated by his past. She stayed up late listening to his stories. She must have known that she was watching a performance, but it was easy to see why "the Greek" appealed to her. She liked his humor: "Of course I'm romantic. You have no idea how romantic it is to make a million dollars," he answered one of her mildly flirtatious questions. She was intrigued by his frankness: "I lie when I have to. Lies are often a lifeline. To survive one must lie." According to Stas, she remonstrated with him only when he told her that the only good politician is a bought politician. Whether Lee knew it or not, the cruise was the turning point in her relationship with Ari. Certainly, anyone who knew Jackie well would have recognized the danger signs as her interest in Ari continued to deepen.

"She wasn't sexually attracted to men unless they were dangerous like old Black Jack," Chuck Spalding had once observed. "It was one of those terribly obvious Freudian situations. We all talked about it—even Jack, who didn't particularly go for Freud but said that Jackie had a 'father crush.' What was surprising was that Jackie, who was so intelligent in other things, didn't seem to have a clue about this one."[10]

Although she was familiar with powerful men, Onassis was something else: His power came from within himself. He was not a figurehead, as Jack was; nor was it his father's bankroll or the ballot box that empowered him. Onassis had made his own fortune. Everything about him was irrelevant before the fact of his wealth. If it is true that wealth is the ultimate aphrodisiac for women like Jackie, his was a lifetime supply.

He told her about his father, Socrates—"a Turkish citizen with a Greek soul"—who had emerged from the interior of Asia Minor in the nineteenth century and made a fortune in the tobacco business, only to lose every penny in the Smyrna massacre of 1922. During the killings, three of Onassis's uncles were hanged; a favorite aunt, Maria, her husband, Chrysostomos Konialidis, and their daughter were burned to death when the Turks set fire to a church in Thyatira in which five hundred Christians were seeking sanctuary. Ari told her how he had saved his own skin by making himself useful to a young Turkish lieutenant who had commandeered the family home. As usual, he made no attempt to hide the true nature of their friendship. He had been young, highly sexed, and his preference was for women, but he was also practical about such matters, and the relationship with the young Turk was as welcome as it was expedient. He told her about his escape from Smyrna, with the family savings bandaged to his body, hitching a ride aboard a U.S. destroyer to the Aegean island of Lesbos, which had become a giant refugee camp; how he had gotten his stepmother and his sisters to Athens, and spent the last of the family money on bribes to get his father out of jail; and how, instead of gratitude, there had been accusations about where the money had gone. He told her about his decision, at the age of seventeen, to emigrate, and how, traveling on a Nansen refugee permit, valid one-way to a country of resettlement, and with $250 in his pocket, he had made his way to Argentina.*

Although Jackie was usually secretive about the details of her own life, she knew how to draw people to her with morsels of juicy scandal about others. She "would be sitting with some old guy who'd almost have nodded off and suddenly ask a question so filled with implied indiscretion that this old guy's eyes would almost pop out of his head," said one Kennedy family friend.[11]

More bluntly, Onassis told Gratsos: "She loves talking dirty."[12]

Jackie had told Ari about her father-in-law's affair with Gloria Swanson in the thirties. Joe Kennedy had regaled her with "intimate details of Swanson's body, particularly her genitalia, making fun of Swanson because, he claimed, she was sexually insatiable, having orgasms not once but five times a night."[13]

Onassis, who had not revealed his own affair with Swanson, asked what Jackie thought Swanson saw in Joe Kennedy. Was it his money? Jackie said that money was important but that an intelligent woman would want more than money. Swanson had needed Joe to help her

---

* Onassis claimed that he was seventeen when he left for Buenos Aires in 1923. He also often added that he was "on the edge of manhood," which is usually regarded to be twenty-three in Asia Minor, especially among families from the interior, where his family had its roots. J.K. Campbell. In his study of moral values in a Greek mountain community, writes, "As a son reaches the threshold of full manhood, twenty-three years of age, the importance of his prestige ranking within (his peer group) becomes incompatible with undue deference to any senior person, even his father" (*Honour, Family and Patronage.* Oxford: Clarendon Press, 1964).

break free of the studio system, she said. But she had done well in the studio system, Onassis pointed out; why would she want to break free of such a comfortable life? Perhaps because it only looked comfortable if you weren't imprisoned in it, Jackie had answered.[14]

It is hard to tell whether these exchanges contained the seeds of something deeper, or if they were simply flirtatious small talk, but it began to cross Artemis's mind that a change was occurring in her brother's relationship with Lee's sister. Years later Artemis would claim that their conversation had not been about Gloria Swanson and Joe Kennedy at all—but about themselves.[15]

Certainly, in speaking to her the way he did, Onassis had succeeded in breaking down the barrier that Jackie had put around herself; he was treating her not as the First Lady, the most celebrated and admired woman in the world, but simply as a woman. He had, said the astute and worldly Irene Galitzine, brought her into his life in a way that he had never done with Lee.[16]

But none of these undercurrents altered the surface of things. "There are maids who bring your breakfast, wash and/or iron your clothes or dry clean if desired," Suzanne Roosevelt wrote a friend. "There is even an operating room somewhere though I haven't seen it. . . ."

They were sailing due northeast, the sea smooth as the silk of one of Princess Galitzine's two hundred dollar pajama suits; islands floated by like faint shadows in the soft October light. People who remember the *Christina* at such moments remember the gentle creaking of the deck, the sound of water breaking at the bow—and, in actor Richard Burton's descriptive phrase, "the saline smell of ocean and sex."[17]

In Washington, a Republican congressman was lambasting the cruise, impugning the integrity of the president's wife, and questioning Roosevelt's motives in accepting hospitality from a foreigner who had plenty to gain from the undersecretary's influence with the United States Maritime Administration. It was Bobby's worst nightmare come true. According to Evelyn Lincoln, Bobby told the president that he must recall the First Lady immediately and "must not take no for an answer."[18]

"But wouldn't that admit the critics were right?" Jackie protested when her husband issued his ultimatum over a ship-to-shore telephone that required a great deal of shouting at both ends. Changing her plans would only give the papers another headline. "Why emphasize something best ignored?", she argued as sweetly as she possibly could while speaking at the top of her voice.

But to one White House insider, "Jackie was making the president pay for all his screwing around."[19] If she refused to come home, what could he do about it? The plain truth was: His own behavior had made him a lame-duck husband.

In Paris, Costa Gratsos and his wife Anastasia were trying to comfort Maria Callas. Pictures of the cruise, of Onassis and Jackie strolling side-by-side through the tiny streets of his past, were appearing every day in the French papers. "Four years ago," Callas said, "that was me by his side being seduced by the story of his life." It brought home to her how much more she had lost in leaving Meneghini than his devotion.[20]

Gratsos was fond of Callas. He knew she could be foolish, egotistical, and often a pain in the ass, but he was saddened by the way Onassis had treated her, and he and Anastasia had refused to join the cruise when they heard that she had not been invited. Nevertheless, Callas had taken the news bravely. "It's not difficult to be swept off one's feet. Living with the consequences, that's the hard part," she told Gratsos.[21] In operas she had played heroines who died for love, and that was something she understood. But a few days later, she miscarried the child that she had hoped would bring Onassis back to her.*

There had always been a sense of performance in Jackie and Lee's relationship, how they would whisper together and break into peals of mys-

---

* According to Nicholas Gage's *Greek Fire: The Love Affair of Maria Callas and Aristotle Onassis* (London: Sidgwick & Jackson, 2000), Callas had a child by Onassis three years earlier, one year after the start of their affair. The baby boy, claims Gage, was born in Milan and died the same day.

terious laughter. But few family conflicts are as cruel and unforgiving as those between sisters. Lee's friend Truman Capote, according to his biographer Gerald Clarke, saw her as "a modern Becky Sharp for whom fate had chosen an exquisitely poignant torture: her childhood rival, her sister, Jackie, had grown up to be the wife of a President and the most celebrated and admired woman in the world."[22] Jackie had always known how to wound while seemingly at her sweetest, and Lee had schooled herself to give no sign of what she suffered in return; it would simply have added to the humiliation if she had.

But it would have been astonishing had Lee not been disturbed at the tone and the direction of the conversations between her lover and Jackie. Her sister's *faux naïf* questions and the rapt attention she paid when Onassis spoke must have been like a replay of so many painful scenes from childhood, when they had been in contest for their father's love and attention. Lee had lost then, and she was losing now. Onassis was simply the latest in the line of older men who had taken Black Jack's place in her life, and Jacks was moving in on him with the sweet madonna smile she always wore when she was doing something unconscionable. It broke Stas's heart, he later told a friend, to see his wife smiling bravely while "twisting her napkin in her lap as if secretly wringing the neck of her favorite rag doll."

Stas knew Onassis well and guessed where it was going, even when it was still only a thought in Onassis's mind. Before he left the cruise at Istanbul, he reminded his wife of how closely the situation developing aboard the *Christina* approximated her own experience with Onassis before he tossed Callas aside for her. Although she must have been nursing her own suspicions, Lee told her husband that he was mad. Quite mad. Her sister, she reminded him icily, was the First Lady of the United States.[23]

A few days later, long-lens photographs of Jackie sunbathing in a bikini on the deck of the *Christina* appeared on front pages of newspapers around the world. Bobby's reaction was more than simple anger, reckoned London *Sunday Times* Washington correspondent Henry Brandon, a friend of the Kennedys. At one level, Bobby felt the fury of the political manager who recognized the damage the publicity was doing to the Kennedy image; at a deeper and more complex level, according to Brandon, was the fact that "Bobby was almost certainly already half in love with Jackie himself, and so you also had the resentment of a frustrated lover."[24]

"There was a Rubicon and Bobby believed that Onassis had sailed straight across it," said Roswell Gilpatric, who would himself become one of Jackie's lovers, and who was at Hickory Hill when the President told Bobby that Jackie had refused to cut short the cruise and return home. Hate, Gilpatric said, was not too strong a word for the feelings the Kennedys felt toward Onassis at that point.[25]

The President's popularity had reached a new low in that week's Gallup Poll; an editorial in the *Boston Globe* asked, "Does this sort of behavior seem fitting for a woman in mourning?" Yet another congressman had accused Jackie of defying propriety by accepting the "lavish hospitality of a man who has defrauded the American public."[26]

But was it simply his brother's presidency that Bobby Kennedy wanted to protect? For as Drew Pearson would later write: "Though the thirteen colonies discarded the divine right of kings a hundred years before Bobby's Irish ancestors migrated to Boston, he still operated on the theory that the nearest relative should inherit the White House throne."[27] And Bobby must have known that it was his future as much as the president's that was being threatened in those distant Greek islands in the autumn of 1963.

The *Christina* had reached the Ionian Sea, where the bright waters of the Aegean flow into Homer's "wine-dark sea," and where Onassis was creating his own private kingdom on the island of Skorpios. Jackie must have felt a million miles away from Washington, American politics, newspaper critics, and her husband's bossy family.

Stripped to the waist, Onassis had toiled on his knees alongside his workers, planting all the trees and shrubs of the Bible. As they walked across his new domain, he told Jackie the stories his grandmother had

told him when he was a child: The fig is the first of the fruits to be mentioned in the Bible; when a pine cone is cut lengthwise, the mark on its
surface resembles the hand of Christ—it was a sign of His blessing on
the tree that sheltered the Virgin Mary when she was fleeing Herod's
troops with her family.

Now, Jackie saw him in a different light; he was nothing like the man
she had met in the South of France a decade earlier. On his island, she
would later tell Artemis, he revealed a depth and a side of himself that
surprised and touched her deeply.

The following day they sailed to Ithaca. It was there—in the principal stateroom named after the storied island, and where Onassis had
spoon-fed Churchill, seduced Garbo, and convinced Callas to leave her
husband—that they became lovers.

According to Onassis, Jackie's susceptibility at that moment was considerable, especially in the context of her hurt at Jack's continuing unfaithfulness. She had recently lost a baby that she must have suspected Jack
planned out of political imperatives. And he had begun a new affair with
a woman who was much more of an embarrassment to her than Monroe
had ever been. For Mary Pinchot Meyer—the divorced wife of CIA chief,
Cord Meyer; niece of the conservationist and two-term governor of Penn-
sylvania, Gifford Pinchot; and the sister of one of her closest Washington
friends, Antoinette (Tony) Bradlee, wife of *Newsweek*'s Washington correspondent, Ben Bradlee—was a woman with whom Jackie mixed socially.
It was this affair, believed one White House insider, that was the final
straw that persuaded Jackie to continue with the cruise despite her husband's objections and pleas to cut it short.

Artemis, who knew her brother so well, would later tell his daughter
Christina that her father had "seduced the President's wife, not his
widow."[28] The look on Jackie's face the following morning, said Artemis,
was proof enough for her.

TWELVE

"SHE EXPECTED THE SUN TO STAND STILL FOR HER"

Many are the false tales recited about women, they are but a weak compensation for the true ones of which we are unaware.

—SENAC DE MEILHAN, 1736–1803

If Jackie's decision to accept Onassis's invitation to join the *Christina* that autumn affirmed that there would be a relationship between them, its consummation at Ithaca determined what sort of relationship it would be.

On the afternoon they became lovers, Onassis later told Georgakis, he had noticed Jackie's jewel case open on her dressing table. Surprised at how few pieces she had, and of how little value they appeared to be, he called Van Cleef and Arpels in Paris and told them to fly a suitably impressive gift to the yacht. They responded with an $80,000 gold and ruby bracelet.[1]

And just as the discovery of the bracelet he had bought for Lee gave Maria Callas her first intimation that Lee had taken her place, so Lee knew that it was now all over for her, too.

"The Princess will get over it," Onassis told Georgakis.[2] Although he dismissed his women ruthlessly when they had outlived their usefulness, his parting largesse usually made the pain of breaking up bearable. According to Alastair Forbes, an old friend of the Kennedys, Lee was "fobbed off with a million dollars"[3]; it may also have been accompanied by a parcel of land on a promontory near Athens that Truman Capote claimed Onassis gave Lee when he married Jackie.[4]

Gratsos, however, worried not about Lee but about what price the Kennedy brothers might yet exact for the cruise. The Haiti deal was still what mattered most to Gratsos, and he knew that Onassis's triumph could ruin everything. As Gratsos told Georgakis: "I kept thinking: this could be the most expensive fuck in the history of the world."

Jackie had known from the start the political damage the cruise would do to the family. But it was not until she got back to Washington that the harm she had done to her own reputation came home to her. Although she was prepared for the embarrassment the episode would cause her husband, she was shaken to discover the extent to which it had undone the sympathy and goodwill she had personally received from the American people following infant Patrick's death.

She had assumed that her place in American hearts was unassailable. But the same newspapers and television channels that only a few weeks earlier had admired her dignity and courage in the face of personal misfortune were now attacking her for her self-indulgence, extravagance, poor judgment, and failings as a wife and mother. It was the most devastating public criticism she had ever received; for the first time since Jack entered the White House, she had become a target for the Republicans and a definite political liability. Although she managed to maintain a semblance of composure—her sangfroid had always been her best defense—she must have been ruffled.

The contrast of her relaxed behavior in Greece to her aloof attitude at home had also irritated the White House journalists, who had grown accustomed to her constant "constraints and demands."[5] UPI's Merriman Smith sniped: "Touring with her sister, Lee Radziwill, Mrs. Kennedy allows herself to be photographed in positions and poses which she would never permit in the United States . . . she's almost a different person when traveling, as it were, on her own. . . ."[6]

In an effort to counter the critical press, Bobby told his brother that he must put on a show of family unity. When Jackie's plane arrived in Washington on October 17, John Kennedy was waiting at the airport with their children and a beatific expression on his face to provide the media with a photo opportunity to record their reunion.

Within the family, however, resentment against Jackie ran deep. For Ethel and the other Kennedy women, nothing that she did was ever right, but on this occasion she had gone too far. In times of need, the Kennedys had always united behind their leader, and only a year away from what was going to be a tough second-term presidential campaign, they were concentrating their hearts and minds on the challenge. Eunice's husband, Sargent Shriver, had put his own political career on hold to support his brother-in-law; even the wretchedly unhappy Peter Lawford had agreed to postpone his own divorce until after the election.[7]

Bobby was furious at Jackie's refusal to toe the Kennedy line. John Kennedy, on the other hand, was more fatalistic than angry. He knew how to be civilized about such matters, as his brother did not. And although, according to one report, it was the president who "informed Onassis, through one of his aides, that he would not be welcome in the United States until after the 1964 election,"[8] Onassis claimed that it was the attorney general who had instructed Jackie to "tell your Greek boyfriend he won't be coming back here until after Jack's reelected . . . a fucking long time after, like maybe never."[9]

There is little doubt that the relationship between John Kennedy and his wife had been deteriorating steadily throughout the presidency, although the truth about the state of their marriage had been successfully hidden from the public. Moreover—until the cruise—no one had done more to sustain the deception than Jackie herself. Her way of remembering this crisis, for example, was the precise opposite of the truth—and a tribute to her own guile as well as the Kennedy's gift for PR. William Manchester, hand-picked by the family to write the official history of President Kennedy's assassination, reported Jackie's version of the events following the death of her infant son. "She had wanted to stay with [the president] and the children. Normality, routine—it seemed the best way to cure her depression," he wrote, claiming that the president "had a different plan: she should forget herself in other lands. Politically, she suggested, this was unwise. With the election a year away, a cruise on a Greek millionaire's yacht was no prescription for votes, but [the president's] mind was made up. So she went . . ."[10]

The White House had never been Jackie's favorite residence, and the regularity and predictability of her absences had enabled her husband to pursue his assignations behind the closed doors of the family quarters without hindrance. Indeed, no other president of the United States can have contrived to invent a life as recklessly promiscuous as his had become by the autumn of 1963.

Nevertheless, he could get away with it only as long as he was able to purchase Jackie's acquiescence—and as long as his father and Bobby were able to continue to sweep his indiscretions under the carpet with "payoffs, legal action, and other kinds of threats to silence women who . . . threatened to go public."[11]

Meanwhile, on her return from Greece, Jackie "remained only briefly at the White House, long enough to argue with her husband—before heading off to her new weekend retreat at Atoka," their thirty-acre estate in Middleburg, Virginia.[12]

She would stay there for two weeks, during which time Dr. John Walsh, her obstetrician, examined her and pronounced that "her recovery was already complete."[13]

The Onassis cruise had sailed dangerously close to exposing the license that Jackie had been discreetly enjoying since her affair with William Holden seven years before. (According to Roswell Gilpatric, she frequently got "guarded invitations" from admirers who knew of the president's permissive attitude toward her friendships with other men.*) And

---

* Gilpatric would later tell friends that he himself had sent her one of those "guarded invitations," together with a book of love poems. Whether he was successful at that

it was almost certainly the reason why she and Bobby had lied to William Manchester that she had returned from Greece "in far better spirits than she had thought possible." [14]

In fact, the atmosphere in the few hours she had spent at the White House before hurrying off to Atoka was, according to Evelyn Lincoln, "unpleasant, unpleasant; I would say very strained." [15] Jackie's private secretary, Mary Gallagher, also seems to have been in little doubt about the tension in the air the day Jackie got back from Greece. Gallagher was convinced from "the clues she saw" [16] that—in her telling phrase—"a significant relationship had been established" between Jackie and Onassis on the cruise. [17]

Nevertheless, in the autumn of 1963, Jackie had more than one reason to want to punish her husband. First, there was his affair with Marilyn Monroe, which, she would later tell Onassis, continued to hang around "like a cheap perfume" long after Marilyn was dead. [18] Another reason was his refusal to end his affair with Mary Pinchot Meyer, who, as a friend and social equal, was more of an affront to Jackie's self-esteem than Monroe, Judith Campbell, or any of his other women ever had been.

But the third reason was even closer to home, and it was Jack's decision to take Lee to Berlin on his historic European tour in June when Jackie was pregnant and unable to travel. Perhaps unaware of Bobby's sly hand in the matter—his determination to get Lee off Onassis's yacht, and to stop the rumors about Onassis's ambition to become a Kennedy brother-in-law—Jackie attributed the invitation entirely to Jack's selfishness. He knew how much she loved state trips abroad, and his failure to understand her disappointment—and to make it all the more unbearable by choosing *Lee* to be her stand-in (did he not understand the first thing about sibling rivalry!)—simply exacerbated the gulf between them.

---

time is unclear. Her "intimate, graceful letter" thanking him for the book also mocked the idea that a gift of such rare sensitivity might have come from "Antonio Celebrezze or Dean Rusk." (David Halberstam, *The Best and the Brightest*, p. 36. London: Barrie & Jenkins, 1972.)

The New York yachtswoman and photographer Hélène Gaillet remembers being on Skorpios with Onassis one evening in 1974 when the question of Jackie's relationship with her sister came up. Recalling the Berlin incident, Onassis said that Jackie had been "prepared to let Lee take her place in her husband's bed, but not her place in history." [19]

But whatever the true state of Jackie and Jack's marriage at this point—ambivalent and complicated as it always was—only the most obtuse, or blasé, of husbands would have been oblivious to the implications of the bracelet of diamonds and rubies Jackie wore on her wrist when she returned from the cruise. And no matter how much Jack may have lost his way in other areas of their marriage, he surely needed no more evidence than that bracelet to realize how well "the Greek"—whose "total understanding of women," according to Maria Callas, "came out of a Van Cleef and Arpels catalogue" [20]—already knew his wife.

But if the president was able to take the situation in his stride—the bracelet, like the cruise itself, troubled him only in its political dimension, believed Lincoln—Bobby was unable to hear Onassis's name without losing his temper. [21] Was this perhaps because Bobby knew something the president did not? Did he, as Onassis believed, have a spy on the *Christina*? [22] There is no evidence that Bobby was getting any more intelligence about what was developing between Jackie and Onassis aboard the *Christina* than what he read in the press. Nevertheless, Onassis was adamant. "He knows I've had Jackie," he told Gratsos. [23]

If this were true, who could the spy have been? Would Roosevelt have admitted that he had failed to do the one thing Bobby had sent him to do—keep Jackie out of the clutches of "the Greek"?

Stas Radziwill was a notorious gossip, but, again, how could he have blown the whistle on Jackie without incurring Bobby's wrath? His closeness to the Kennedys was as valuable to him as his defunct title. Anyway, if Onassis is to be believed, Jackie surrendered to his charms *after* Stas had left the yacht at Istanbul.

Then could it have been Lee? She had lost no time in telling Jack which way the wind had blown in the Aegean that fall. William Manchester would later interpret her letter as "a feigned sulk," and her complaint that Jackie had been laden with presents from Onassis, while she herself got only "three dinky little bracelets that Caroline wouldn't wear to her own birthday party," as sibling banter.[24]

But it was far from feigned: Her anger was all too human. *"I can't stand it,"* she wrote the president in anguish when the terrible consequences of her lover's change of heart had begun to dawn on her.[25]

Even so, all this would still have remained largely a matter of hearsay and conjecture but for Evelyn Lincoln. Although the least glamorous and most self-effacing courtier of Camelot, Lincoln was one of the most trusted members of JFK's White House entourage. They had been together from the start of his Senate career in 1953 until November 22, 1963, when she was riding behind his limousine in Dallas.

There was not much about JFK that Lincoln did not know; after his death, she instinctively spoke from his perspective. And her opinion that Jackie began her affair with Onassis *"before Dallas,"* almost certainly reflected the President's own suspicions.[26]

"Toward the end," Lincoln said unequivocally in 1995, speaking with an energy and a purpose that belied her own approaching death, "Jackie saw other men, too. You know, the jetset, they have a life of their own. They think nothing of having outside love affairs—outside of their marriage—and Jackie was more or less of the jetset . . . Onassis became more than just a friend."[27]

JFK, she said, had warned Jackie before the cruise that if she accepted Onassis's invitation, the consequences could be incalculable. "He told Mrs. Kennedy that a man like Onassis always wanted something. He said it would be a terrible mistake to accept Onassis's hospitality because he would eventually want something in return," Lincoln told me.[28]

Jackie had refused to listen; she wanted to go on the cruise and nothing was going to stop her. "When Mrs. Kennedy was in that kind of mood," Lincoln said, "she expected the sun to stand still for her."[29]

The fact that Jackie's strength derived directly from her husband's weakness had been one of the most closely held secrets in the White House. But had Dallas not happened, Lincoln told me: "The cruise would have been seen as a catastrophic wrong turning . . . it would have

A favorite port of call for Onassis was the island of Capri. "Apart from Skorpios, this is the island of seduction. If a man cannot seduce a woman on Capri, he is not a man," Onassis would tell friends. Jackie sent *her* friends postcards praising the ruins of the island's medieval castles.

Onassis shortly after arriving in New York in the summer of 1940. Down to his last $2.5 million, but happy to have escaped the war in Europe, he wrote to his Norwegian mistress, Ingeborg Dedichen: "<u> </u> very much what will happen." AUTHOR'S COLLECTION

A rare moment of family togetherness: Neither Ari Onassis nor his first wife Tina took parenthood seriously. Their children, Christina and Alexander, lived in a charmed world of pampered neglect: their parents' affection scrawled on postcards from distant places and wrapped in secondhand hugs from passing friends and strangers. GETTY IMAGES/HULTON ARCHIVE

Tina Onassis's love for her sister, Eugenie, did not stop her from sleeping with (and later marrying) Eugenie's husband Stavros Niarchos, who was also Aristotle Onassis's archrival.

Stavros and Eugenie Niarchos: Did he murder her—or had she simply had enough of his infidelities?
TOM BLAU/CAMERA PRESS LONDON

Before his untimely death, Aristotle Onassis's son Alexander had a passionate love affair with Fiona Thyssen, a woman sixteen years older than he. She said of their relationship: "It took me a long time to stop fighting and accept that we had become indispensable to each other and should just try to

Princess Grace despised Onassis. It rankled her that J. Edgar Hoover called Ari "the real ruler of Monaco." BETTMANN/CORBIS

"He's a great kid—he hates the same way I do," Joe Kennedy said admiringly of his son Bobby.

Twenty-four hours before the death of President Kennedy, Jackie stands by her husband in a bid to make amends for the political damage her cruise with Onassis had caused Jack.

Callas's affair with Ari did not bring her the happiness she had yearned for: "When slight has followed slight, and insult has been added to insult, the love which remains is often illogical, but it is also indestructible," she said. GETTY IMAGES/HULTON ARCHIVE

"The Kennedys could accept me as Lee's lover: that was personal. What they couldn't accept was the idea that I might actually marry her: that was politics," said Onassis of Jackie's sister, Lee Radziwill, before Jackie came on the scene. GETTY IMAGES/HULTON ARCHIVE

...orst nightmare comes true. Jackie knew that pictures of her and Lee... ...the Aegean in October 1963 would cause a political furor in Washin... ...res did to her own reputation shocked her deeply. AP

...ht of himself as a great lover. "All Greek men beat their women: He w...

The beginning of Camelot: John F. Kennedy's funeral. Jackie accompanied by Bobby and Ted Kennedy.

Bobby was always there to comfort Jackie in her times of sorrow. This photo was taken in Hyannisport, Massachusetts, several months after JFK's death.

Bobby Kennedy's assassin, Sirhan Sirhan. Witnesses recalled Sirhan's "enormously peaceful" eyes after the shooting. BETTMANN/CORBIS

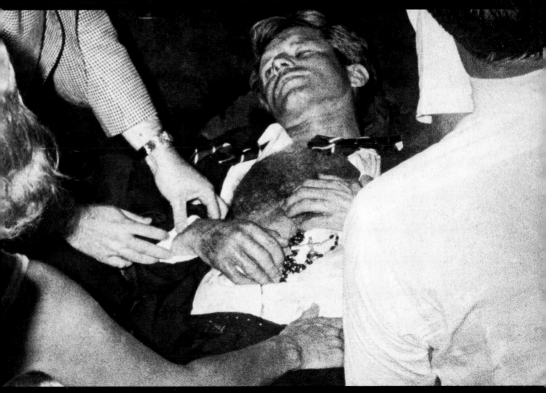

Bobby Kennedy just after the shooting, June 5, 1968. "I guess the kid had everything but the luck."

Jackie at Bobby Kennedy's funeral. She loved Bobby and said of him: "He's the only one I'd put my hand in the fire for." BETTMANN/CORBIS

Onassis driving Jackie and his new stepdaughter back to his yacht after their wedding on his island, Skorpios. On paper, the marriage lasted six years; in reality, it was over within weeks.

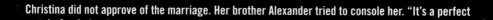

Christina did not approve of the marriage. Her brother Alexander tried to console her. "It's a perfect

Yannis Georgakis *(left)*, one of Ari's closest confidantes, said of the tycoon, "Ari is a charming psychopath ... with no moral imperatives at all." CAMERA PRESS LONDON

Onassis invited Hélène Gaillet to his island, Skorpios. "When I lived at that level I accepted things, I knew things, and did things, that now frankly amaze me," she recalled. HÉLÈNE DE GAILLET

"The good stuff is what gets you killed, kid," Johnny Meyer told his putative biographer, Brian Wells. Meyer was also part of Onassis's circle of close confidantes and business colleagues. BRIAN G. WELLS

"There was always something he didn't want you to know about when you were dealing with David Karr," said Lord Forte. Karr was one of Onassis's confidantes and the source of many of his questionable connections, including the terrorist Mahmoud Hamshari, with whom Onassis made a deadly bargain. BETTMANN/CORBIS

The fanaticism that defined Mahmoud Hamshari, the hatred that had molded him, were masked in polite society. Just after the Six Day War, Hamshari suggested to Fatah that they kill a high-profile American on American soil. Fatah declined, but Hamshari was undeterred. AFP

"In all my years in the church, I don't recall another funeral where the widow was pushed into the background this way," said Greek Archdeacon Stylianos Pirounakis after Onassis's funeral. BETTMANN/CORBIS

exposed everything."[30] The Kennedys' marriage had reached such a nadir, she told writer Laurence Leamer, that if the president had survived and won the election in November, there would probably have been "the first divorce in the White House."[31]

A hint that at least the parameters of a deal had been worked out between Jackie and Onassis came at a lunch at the Hotel de Paris in Monte Carlo shortly after Onassis returned from the cruise.

Present with Onassis were Costa Gratsos, Yannis Georgakis, Roberto "Tito" Arias (the lawyer who Onassis said "read the fine print and invented the loopholes" for his Panamanian corporations[32]), Onassis's brother-in-law and physician Theodore Garofalides, and the ubiquitous Johnny Meyer.

Although Monaco and the prince continued to be an irritation ("your husband makes me feel like a *gauleiter* holding down a troublesome out-post of my empire," Onassis had told Princess Grace[33]), and the Haiti project was going too slowly for his liking, Onassis was in high spirits. He asked Meyer whether he had ever made love to a princess. When Meyer told him only Jewish ones, he roared with laughter and said, "Johnny, let me tell you, you haven't missed a goddamn thing."[34]

"I knew when he made that crack he was no longer planning to marry Lee," Gratsos later told Georgakis.[35]

And the only reason why he would pass up Lee, Gratsos was con-vinced, was because Ari had decided to marry her sister instead.

In Washington, Jackie had changed her tune following her sulky sojourn at Atoka. She had become far more compliant and agreeable. Although she found football boring, according to William Manchester, she was thinking of attending the Army–Navy game with the president after Thanksgiving. "We'll just campaign," Manchester reported her telling Jack. "I'll campaign with you anywhere you want."[36]

"Maybe now you'll come with us to Texas next month," Jack later said to his wife at dinner with Ben and Toni Bradlee.

"Sure I will, Jack," she told him, not missing a beat.[37]

THIRTEEN

THE GREEK WAY

Too heavy a price may be paid
for wealth.

—ST. MARC GIRARDIN, 1801–1873

On Friday, November 22, 1963, the news from Dallas came like a clap of thunder reverberating around the world. At a cocktail party in Hamburg to celebrate the launch of his new tanker *Olympic Chivalry*, Onassis was told the news only minutes after Bobby Kennedy, eating lunch at his home in Virginia, heard about it from J. Edgar Hoover.

The assassination of a president is one of the most traumatic single acts, short of war, that can be inflicted on the psyche of the American people. But the assassination of John F. Kennedy, a man many believed was the greatest U.S. president of his century, stunned and moved the entire world. His youth—or the appearance of youthfulness—his vigor, charisma, beautiful young wife, and small children were part of his charm and popularity. Radio and television networks around the globe preempted their regular schedules and canceled commercial advertising for unprecedented saturation live coverage of the tragedy; in America, it went on continuously for four days and nights.

Unable to reach Jackie—so many people picked up telephones when they heard the first bulletins that nearly every major exchange in Washington went down[1]—Onassis eventually got through to Lee in London late that evening. She asked him to accompany her and Stas to Washington for the funeral. He reminded her that he had been warned not to step foot inside America for at least a year. "I don't think that matters very much now," she told him.[2]

Lee was too smart to make such an offer off her own bat. Truman Capote—who was, at that moment, still one of Lee's dearest friends,* and a confidant with whom she could discuss "the most serious things about life and emotional questions"[3]—later told Joe Fox (our mutual editor at Random House), that the invitation had actually been Jackie's idea.

The First Lady knew that she could not invite "the Greek" herself, Capote claimed, but by having Lee include him in her personal party, he automatically got one of the family's invitations to stay at the White House. "Lee was deeply in love with the Golden Greek but she was a cunt; she played right into Jackie's hands," Capote told Fox.[4]

Although he was one of only half a dozen people outside the family to be given the honor of staying at the White House, Onassis's presence went almost unnoticed in the days of shock and mourning that gripped the nation. On Sunday evening, Rose Kennedy dined upstairs with Stas Radziwill; Jackie and Lee were served in the sitting room with Bobby. Onassis, together with the other house guests, ate with the rest of the Kennedys in the family dining room. At about 10:00 P.M., Bobby Kennedy joined them for coffee.[5]

This was the first time he had come face-to-face with "the Greek" that weekend. Whether Jackie had told him that Onassis was included on the family guest list is unclear, although Bobby would certainly have

---

* The friendship ended in tears when Lee testified on behalf of Gore Vidal in his libel action against Capote, who claimed that Vidal had been kicked out of a Kennedy White House party for being drunk. Lee told Liz Smith that she was "tired of Truman riding on my coat-tails to fame." Anyway, what difference did it make, because they were "just a couple of fags." (*New York Daily News,* p. 3. September 23, 1984.)

known that the invitation could not have been issued without her approval. Was Onassis's presence her way of letting the Kennedys know that she no longer expected to be treated as "just a thing, just a sort of asset, like Rhode Island"?[6] Or, more significantly, perhaps, to remind Bobby that she now held the key to his political fortunes?

While pregnant in 1960, Jackie had played little part in her husband's presidential campaign. "Occupying the throne suited her better than fighting to win it," Barbara Kellerman had pointed out in her book *All the President's Kin.*[7] Nevertheless, Jackie's aloof style—which some people mistook for a kind of spirituality—became one of the most powerful assets John Kennedy possessed. But those who knew her well were aware that behind the image she was something else. According to *Look*'s Stanley Tretick, the Kennedys' favorite photographer, she had a way with her that "sort of strikes terror to your heart."[8]

It was this side of Jackie that must have given Bobby pause at that time. For in spite of his authority as attorney general, his prerogatives as Jack's closest brother and political heir, his rage, and whatever protocol and good taste seemed to demand, he knew that there was no way he could eject Onassis from the White House mourning party without alienating Jackie.

The myth of Camelot had not yet been created, of course. But the idea was palpably in the air.* Certainly it must have been evident to

---

* Trying to write an epitaph of the Kennedy administration, one week after the assassination, in the presence of the president's widow, T. H. White would recall: "At 2 A.M. I was dictating the story from the Kennedy kitchen to two of my favorite editors . . . who, as good editors, despite a ballooning overtime printing bill, were nonetheless trying to edit and change phrases as I dictated. Maness [one of the editors] observed that maybe I had too much of "Camelot" in the dispatch. Mrs. Kennedy had come in at that moment; she overheard the editor trying to edit me, who had already so heavily edited her. She shook her head. She wanted Camelot to top the story. Camelot, heroes, fairy tales, legends were what history was all about. Maness caught the tone in my reply as I insisted this had to be done as Camelot . . . So the epitaph on the Kennedy Administration became Camelot—a magic moment in American history . . . which, of course, is a misreading of history. The magic Camelot of John F. Kennedy never existed. . . ." (Theodore H. White, *In Search of History: A Personal Adventure,* p. 524. New York: Harper & Row, 1978.)

Bobby that part of his future presidential appeal would depend on the power of his brother's legend—and Jackie's role in it. Bobby knew the score exactly. Caught in the crosshairs of grief and ambition, his opportunism did not fail him: Allowing Jackie to bring her lover to Jack's funeral was a small price to pay for her political support in the future.

Indeed, although some noticed that his eyes were not the smiling ones the Irish sing about, Bobby's restraint when he met "the Greek" that Sunday evening at the White House could have been a master class for budding politicians who don't want to be inhibited by moral conscience. Nevertheless, as the booze flowed, he began to taunt Onassis about his wealth and its source, his yacht, his airline, his principality, his past.

Onassis knew that when people did not like him they never failed to inquire about his past. According to William Manchester's official account, Bobby "badgered him mercilessly about his . . . Man of Mystery aura," then suddenly left the room and returned with a contract "stipulating that Onassis give half his wealth to help the poor in Latin America." The document was "preposterous (and obviously unenforceable)," and the Greek millionaire signed it in Greek.[9]

Just as Manchester had previously misconstrued Lee Radziwill's anger over Onassis's "dinky" gift at the end of the cruise as simple sibling rivalry, the Kennedy scribe also reported the incident as nothing more than affectionate ribbing between friends intended to "provide comic relief of sorts"[10] and all part of the "lunacy (that) was creeping into the funeral preparations."[11]

This was not how Onassis remembered it. Bobby was caught in the deepest throes of shame, anger, and frustration at having to entertain the man who had cuckolded his brother. And whatever the others present that evening believed, or pretended to believe, or were simply too drunk or stoned to understand, Onassis knew that Bobby was not baiting him simply to relieve the tension of the moment. Bobby's mockery hid a far deeper hatred. "Bobby did everything he could to humiliate me tonight but I didn't take the bait . . . the more I smiled, the madder he got," Onassis told Gratsos on the telephone later that evening.[12]

*   *   *

On December 3, Onassis was back in Paris for Maria's fortieth birthday party. The coolness between them and the spasm of contempt Maria had felt for him since her exclusion from Jackie's cruise in October seemed forgotten as they sat together at their favorite table at Maxim's, surrounded by old friends. Although one guest felt that Onassis was behaving like "a pimp treating his woman," it was a feather in Callas's cap to have gotten him back for such a publicized event in their hometown.

In a telephone conversation the following day, Jackie told Onassis that since their relationship would not "escape the meaner attention of our friends"[13]—Onassis suspected she meant Lee as well as Bobby and the rest of the Kennedys—she thought that it would be wise for them not to be seen together until after her official period of mourning ended in the summer. Onassis was happy to go along with this. He needed to turn his mind to business again and to the problems of Monaco.

The principality had incurred the wrath of French President Charles de Gaulle, who wanted to put a stop to the thousands of French companies that registered there in order to avoid French taxes. Rumors that he planned to abolish the principality's independence had triggered a slump in the property boom, and as a result Rainier pressured Onassis to plow his SBM profits into new hotels.

The surprising beneficiary of this contest was Callas. With Onassis's time now divided between Paris and the South of France, and Jackie still living in Washington, Callas was the one to whom he again turned for comfort. Friends hoped that he had finally realized how much he needed her; some were also dismayed at how much of herself she was still willing to sacrifice for him. It must have violated everything in her nature to fly to Port-au-Prince to sing for "Papa Doc" Duvalier at his presidential palace when he began to prevaricate over his development deal with Onassis.

In his perverse way, Onassis liked Papa Doc, whom he believed would not "dare exceed the bounds of reasonable treachery with me."*

---

* His confidence proved unfounded. Onassis's problems in Haiti began when another player slipped into town: Sheik Mohamed Fayed (now better known as Mr. Mohamed

Their mutual hatred of Washington, the CIA, and particularly the Kennedys, was also a kind of bond. Onassis even attended a voodoo ceremony at the dictator's palace to cast a spell on the Kennedys—but only

---

Al Fayed, chairman of the London store, Harrods, and owner of the Ritz hotel in Paris). Allegedly from Kuwait, Al Fayed was rumored to be the cousin of the emir and was loaded with petro-currency. Papa Doc continued to string Onassis along while he found out what the emissary from Kuwait had to offer. According to one report, Fayed was "kept isolated to prevent enemies of the regime from trying to turn him away from the work which had been entrusted to him, which was to make the emir's millions rain down on the country." But despite these precautions, Onassis found out about Fayed and demanded an explanation. Fayed, he was told, was financing a state automobile insurance scheme. This was true, as far as it went. But even as Duvalier was assuring Onassis that he had nothing to worry about, the text of a decree approving a deal between the government of Haiti and Sheik Mohamed Fayed's newly formed Haitian Petroleum Development Company, S.A., which was identical to the deal he had negotiated with Onassis, was being promulgated in a secret edition of the Haitian official gazette, Le Moniteur. According to U.S. Ambassador Timmons, only four copies of this issue of Le Moniteur were published. Such was the speed and secrecy with which the negotiations were concluded that it was not until a month after the sub rosa proclamation in Le Moniteur that Timmons even got wind of what had happened: "Would appreciate any info available re Mohamed Fayed who recently visited Haiti claiming be wealthy businessman (it also rumored he really Egyptian)," he cabled the State Department on September 26. It is hard to fathom how a man so deeply perceptive in his understanding of greed and deception as Onassis failed to rumble Papa Doc's lies. But despite the rumors of his rich partners in Kuwait, Fayed seemed to be having trouble finding the capital to finance the scheme. On February 14, 1965, the New York Times ran a one-sentence inside page story carrying a Santo Domingo dateline: "Mohammed (sic) Fayed, a foreign promoter who was granted concessions by the Haitian Government, has been reported missing. According to President Francois Duvalier, Mr. Fayed has taken a large sum of money out of Haiti. President Duvalier said Mr. Fayed was being sought throughout the world, and especially in Kuwait, from where Mr. Fayed arrived last June." On February 7, seven days before he announced that his former partner was "being sought throughout the world," the pro-Duvalier Le Nouveau Monde ran an editorial exculpating the government from all blame. It is reported that Fayed was the victim of a "national and international conspiracy" hatched to destroy a deal which could only have strengthened the Haitian economy, and said that Fayed was the "quarry of a veritable hunting pack of businessmen from Haiti and U.S." The "hunting pack," according to Onassis, was the CIA. The Agency had set out to destroy his Haiti deal as it had destroyed Jiddah, and would later drive

after Jackie had told him that Bobby had bought an Onassis voodoo doll which he pierced with needles.*

At the palace ceremony, a vial of air, a morsel of earth, and shreds of flowers from President Kennedy's graveside were used to cast a spell upon the Kennedys.[14] Six weeks later Teddy Kennedy's plane crashed in a thunderstorm; the pilot and an aide were killed; Teddy was seriously injured. When Jackie told Onassis what had happened, he called Gratsos and

---

him out of Monaco, he later claimed. Certainly, Fayed would have been a perfect dupe to ruin Onassis's Haiti deal, with his bewildering background, and his family and close business ties to the pro-American Saudi Arabian billionaire Adnan Khashoggi—codenamed "Dynasty." Khashoggi was one of the CIA's most valued "well-informed private individuals," a phrase used by the Agency to describe a particularly important intelligence source. According to former CIA station chief Miles Copeland, Khashoggi's "Dynasty" reports on the strategic importance of the Arabian Peninsula and the Persian Gulf "were so perceptive . . . CIA analysts suspected they were written by some genius in one of the major oil companies." But if Fayed was being "run" by the CIA, perhaps even without his knowledge—it was possible to finance a proprietary without the front man knowing who was financing the operation, according to Copeland—why had the Agency allowed their man to be hung out to dry? "Why does the phrase 'killing two birds with one stone' come to mind?" said Copeland. Copeland, who became Adnan Khashoggi's consultant and confidant, said that after Fayed left Khashoggi's company and divorced his sister Samira, "there was no love lost between those two men . . . they were the mid-east's version of Onassis and Niarchos." It would not be difficult to imagine a scenario in which Fayed was set up by the CIA to sabotage Onassis's big deal in Haiti, as well as to cause Duvalier maximum grief, and then pull the rug from under his feet. "It would have been a neat way to say thank you to Khashoggi for old favors, and also, when the Fayed deal failed, and with Onassis bounced out of the picture, Duvalier would be forced to negotiate with the Americans. Which he did. We used each other all the time—often with only one or the other of us realizing it," said Copeland.

* Although this story at first seems almost too bizarre for credence, according to *Life* magazine (November 18, 1966) two weeks before JFK's assassination Bobby acquired a Lyndon Johnson voodoo doll. According to Peter Collier and David Horowitz (*The Kennedys: An American Drama* p. 319. New York: Summit Books, 1984) this had been "the source of endless merriment as everyone poked it with needles."

said, "This voodoo business really works—but they got the wrong Kennedy!" [15]

Meanwhile, although President Kennedy's withdrawal of U.S. aid had left Haiti in desperate need of foreign capital, "Papa Doc" continued to keep a firm hold on power and forced President Johnson to reappraise America's relationship with the dictatorship. At the May Day celebrations, six months after President Kennedy's assassination, the U.S. ambassador attended the high mass to praise Duvalier's rule.

"Papa Doc realized immediately that he was confronted by two interests which were by no means compatible," Georgakis later recalled Duvalier's dilemma as he found himself being wooed back into the American fold just as he was about to go into business with perhaps the one man in the world whose name was almost as discredited in Washington as his own had been. [16]

Jackie, meanwhile, was never far from Onassis's mind. They spoke on the telephone several times a week; he sent her discreet gifts, anonymous bouquets, books (including Edith Hamilton's *The Greek Way*), and regular sums of money. "It was like a wartime romance," said Costa Gratsos. [17] And like a wartime romance in which partners grow and change in each other's absence, Jackie had slowly metamorphosed into the Holy Widow. Her haunting presence at JFK's funeral, seen live on television satellite in more homes and more countries than any other single event in history, had turned her into an American icon. Her face was now familiar not only in the capitals of the world, but also in the poorest villages from the Punjab to the Euphrates valley.

By inventing Camelot, she had made herself the most important Kennedy of all. More than the keeper of the flame, she had become the very incarnation of the Kennedy myth on which the family's political destiny depended. She had also become what she had dreamed of being all her life: a superstar.

She attended charity balls, protested against buildings that cast shadows over Central Park, attended gallery openings. But that alone wasn't enough of a life to bring her a measure of contentment. To combat her bouts of depression, she began taking vitamin shots laced with speed, prescribed by John Kennedy's former New York physician Max Jacobson, a.k.a. Doctor Feelgood.*

Onassis told Georgakis (one of the few who knew how close he and Jackie had remained and how much the friendship was costing him): "She's a mess. She doesn't know what she wants. Bobby's fucking up her mind the way he fucked up Monroe's." The bitterness of his tone surprised Georgakis. He knew that Onassis had no time for Bobby. He knew that Ari resented the way Bobby had seized on the chance to ridicule him at his brother's wake at the White House, while surrounded by his cronies, and knowing that Onassis could not defend himself without creating a scene. It had been a cowardly attack, and Onassis had emerged with dignity. But Onassis's remark about Bobby playing mind games with Monroe puzzled Georgakis. What did that have to do with anything? Georgakis asked. Onassis was shocked at his obtusity. Bobby had used Monroe to destroy their friend Spyros Skouras. "Now he is turning Jackie against me," he said gravely.[18]

To make matters more complicated, in mythologizing her marriage to make it tally with the Camelot legend she and Bobby had created, Jackie had not only breathed fresh life into a relationship that was on the point of total collapse a few weeks before Dallas, she had also half fallen in love with the husband she had invented to fit the fantasy.

But if she grieved for a past that never was, it did not mean that her grief was not real. It was as real as Bobby's, who would not say the word *assassination* or *Dallas,* and would refer only to "the events of November 22."[19] Though he had been a more devout Catholic than any of his brothers, Bobby did not have his wife's faith in the church, and had turned to literature and philosophy for consolation: ". . . in the late afternoon before the fire in Jacqueline Kennedy's Georgetown drawing room, in his reading—now more intense than ever before,

---

* Only later was there concern within the medical circle about potential side effects, which included addiction and psychiatric complications (*New York Times,* December 4, 1972). Jacobson lost his medical license in 1975.

as if each next page might contain the essential clue—he was struggling with that fundamental perplexity: whether there was, after all, any sense to the universe," wrote his friend and biographer, Arthur Schlesinger.[20]

Jackie gave him the copy of *The Greek Way*, by Edith Hamilton, that Onassis had given her. Bobby devoured it, scrawling marginalia and marking lines from Aeschylus and Herodotus[21]—themes, wrote Schlesinger, that spoke to his anguish.[22]

FOURTEEN

THE HEART AND MIND OF A CLASSY COCOTTE

One must not judge men by what they don't know, but by what they do know and by their manner of knowing it.

—VAUVENARGUES, 1715–1747

Jackie moved to New York City in 1964. Bobby followed shortly afterward; he planned to launch his campaign to become New York's Democratic senator, his first step toward his run for the presidency.

Ever suspicious, Onassis was convinced that their love affair had become more than epistolary. The tabloid press and Washington society were full of stories about their burgeoning romance. "Jackie and Bobby were as close as you can get. What do I mean by that? Just anything you want to make of it," said Charles Spalding, a friend of the Kennedys.[1]

But no matter how enmeshed Jackie's life with Bobby had become, and even in a society in which duplicity is often part of the game, it would have been madness for her to be sleeping with two hubristic men whose hatred for each other positively seethed.

Johnny Meyer said, "Ari was pissed when he found out that Bobby had followed Jackie to New York. He said, 'Bobby's going to fuck her, surest thing you know. Jack fucked her little sister, Jackie'll fuck his little brother.'[2] Gore Vidal agreed with Onassis's sentiments exactly, although he expressed it rather more elegantly: "I suspect that the one person [Jackie] ever loved, if indeed she was capable of such an emotion, was Bobby Kennedy. As Lee had gone to bed with Jack, symmetry required her to do so with Bobby."[3]

But, however phrased, Onassis could live with it: He had shared women before; he preferred them to be experienced and drawn to a whiff of debauchery in a man. He had taken Tina to sex parties in Paris, sometimes picking up wealthy strangers in the *Bois de Boulogne*. As Ari grew older, he increasingly depended on kinky stuff to excite him. So, when he convinced himself that Jackie was sleeping with Bobby, he was angry but also aroused. "She's no better than Tina," he told Yannis Georgakis, alluding to the fact that Tina had embarked on an affair with *her* brother-in-law, Stavros Niarchos.[4]

But if Jackie's intimacy with Bobby and Onassis pushed them all closer to their destinies, other fates were also at work.

According to historian Arthur Schlesinger, *The Greek Way*, the little book that Jackie had received from Onassis and passed on to Bobby, had helped Bobby to come to terms with the moral crisis that afflicted him after Dallas; it had driven him on toward his great ambition of the U.S. presidency, and Schlesinger claimed it opened up "a world in which man's destiny was to set himself against the gods and, even while knowing the futility of the quest, to press on to meet his tragic fate."[5]

Yet even at this stage, Bobby's romance with Jackie, consummated or not, could have been stopped dead in its tracks; but those whose trust they were betraying seemed to have deliberately averted their attention from what was happening in New York, where, according to one report, Jackie and RFK were "frequently spotted holding hands and whispering in darkened corners of . . . romantic clubs and restaurants."[6]

When Eunice said, "Well, what are you going to do about it? He's spending an awful lot of time with the widder," and Ethel made no reply, her silence said it all.[7] It was not that she was shirking the consequence of recognition, or that she had simply become another Kennedy wife living in fake ignorance about her man. Ethel was not naive. But with all their children, Bobby's Catholicism, and the compulsion of the Kennedy legacy, she knew that her marriage was unassailable. And indeed, perhaps more than Bobby himself, Ethel wanted her husband to run for the presidency in 1968; and however irritated she may have been about what they were getting up to in New York, Jackie remained the best political asset Bobby had. If their affair was the price Ethel had to pay for Jackie's loyalty, so be it.

In addition to the annual $10,000 presidential widow's pension, and the $100,000 or so a year she was getting from the interest on her children's trust fund, Jackie was also collecting $50,000 a year from Bobby. But this was still not remotely what her income needed to be, and by autumn of 1964, Onassis was sending regular donations to enable her to maintain a lifestyle that was commensurate with her standing as the future Mrs. Aristotle Onassis.

She moved from the smart but domestic designer Oleg Cassini to Valentino and the cream of European couturiers, from jewels borrowed from Tiffany's for state occasions[8] to her own Van Cleef and Arpels collection, which Onassis had started so handsomely to celebrate, as he would later indelicately put it, "our first fuck."[9]

According to lawyer Roy Cohn, Onassis told him that Jackie had excited him in those early days because he had always liked whores—and Jackie had "the heart and mind of a classy cocotte."[10] (Like many wealthy Greeks, Onassis had a passion for sex but little real interest in women; he coped with this dilemma by using call girls. "The best girl is the girl you never have to see again," he once told me.) It hardly troubled Onassis that Jackie was sleeping with Bobby, Cohn told me. "He never made the mistake of falling in love with Jackie; he was never jealous of her relationships with other men. But he was a complex man. He could accept the fact that she was sleeping with Bobby Kennedy, but he could not forgive Bobby Kennedy for sleeping with her. I suppose that is very Greek."[11]

According to Onassis, rather than being defensive about her affair with Bobby, Jackie told him stories that she knew would "pique his prurient curiosity," said Georgakis. When Onassis railed against Bobby, calling him "a little prick," she smiled and said that he had that wrong! Nevertheless, when Bunny Mellon gave Jackie an antique bed for her new $200,000, fourteen-room apartment on Fifth Avenue,* Onassis suspected that it was a gift from Bobby and made her get rid of it—but not until after they had spent a night making love in it.

It is unlikely that Maria Callas knew to what extent Onassis had become part of a romantic triangle with Jackie and Bobby, but she was aware by the early spring of 1965 that all was not well with their own relationship. Onassis accompanied her to New York for her two performances of *Tosca* at the Met. In truth, it was simply an opportunity for Onassis to spend some time with Jackie. Her presence at Callas's first night, where— accompanied by one of her several homosexual "walkers"—she appeared magnificently dressed in white mink and diamonds, bestowed an air of royal command on the evening.

Upstaged, her nerves shredded by the charged atmosphere of the occasion, Callas was not at her best. Her high notes sounded shrill and achingly insecure, and one critic reported that she "relied almost wholly on dramatic rather than vocal brilliance to carry her through."[12] She needed Onassis's comfort more than ever, but she returned to Paris alone.

---

* The apartment sold shortly after Jackie's death in 1994 for over $10 million. According to Christina Onassis, her father bought the Fifth Avenue apartment for Jackie in the spring of 1964. Onassis told me that he had contributed to the purchase price: "I helped out a little . . . Bunny Mellon (wife of philanthropist and Kennedy friend Paul Mellon) did, too." Privately, he was less philosophical about the arrangement. "It was no secret in the family that he frequently sent Jackie envelopes filled with cash. He felt he had paid for the apartment twice over. He was furious when he found out that Jackie had also accepted money from other people toward it," said Yannis Georgakis. It was not a question of ownership or property right. "Ari wanted to feel that the apartment was something private between them, and not shared with half a dozen other benefactors."

*   *   *

Jackie's move to New York had been as much Onassis's idea as hers and Bobby's. Washington was not Onassis's favorite town; it was not a town he could slip in and out of without being noticed. New York was more convenient, easier to hide in, and by the late winter of 1965, he was spending as many weekends in New York as he spent in Paris.

Jackie was even more newsworthy now than when she was First Lady. Fans, as well as the *paparazzi,* camped outside her apartment. The "watchers," as she called them, made her nervous, and Onassis—tetchy that he was unable to visit her in the apartment he had generously contributed toward—rented another apartment on East 64th Street for their trysts. "My stamina astonishes her," he told Georgakis. For the first time in her life, she was in the hands of an expert—and eager to learn, he said.[13]

Meanwhile, Jackie's attempts to censor parts of William Manchester's book *The Death of a President,* her authorized history of the assassination of John F. Kennedy, had become front-page news.

Assured by Bobby that there would be no problem in manipulating the historian,[14] she had spoken to him with unusual candor—albeit not without a warning: Unless she ran off with Eddie Fisher, no one would cross her and survive, she told him, still shining with goodness and confident of her place in the hearts of the American people.[15]

But as *Look* magazine prepared to launch its serialization, and Jackie's demands for more and more cuts grew, even Manchester began to balk. Inevitably, details of the stories she wanted removed leaked to the press—there must be no mention of her chain smoking, no hint of her drinking, which had become noticeable, nor of her separate sleeping arrangement with the president on their last night in Texas—and exposed not only the extent of her vanity but also, in Kitty Kelley's telling phrase, "the imperiousness which lay behind that black veil."[16]

But one of Jackie's most vehement demands went almost unnoticed. It was that Manchester destroy every page of his transcripts from the letters she had sent to her husband from Onassis's yacht a few weeks before the assassination.* Since these letters must have revealed a measure of duplicity, as well as moral ambiguity, at the heart of her character, her anxiety to destroy them is understandable.

For Manchester this was almost the last straw. On the edge of nervous collapse, and suicidal, he told his agent: "I have reached the point where, if the integrity of my manuscript is violated, I have no wish to go on living . . ."[17]

It was at this point that Jackie turned to Onassis. Could he help her stop its serialization? He told her that the only way to do so was to buy back the rights to the book, for which *Look* magazine had paid $665,000. Of course, he said, she would have to "sweeten the pot."[18] Apparently under the impression that Onassis had offered to supply the sugar, she summoned *Look* publisher Mike Cowles to Hyannis Port. "I simply don't understand why you should proceed with publication when I don't want you to," she began in an astonishingly high-handed harangue before making her offer: "If it's money, I'll pay you a million."[19]

It wasn't a question of money. *The Death of a President* was an important book that could not simply be bought off the market, Cowles told her, astonished at how little grasp she appeared to have of its historical significance.

Increasingly hysterical,[20] and egged on by Onassis, Jackie turned her spleen on Bobby, whom she accused of being more concerned with how he could duck what had become a political boomerang for him than what private grief it might cause her. She was not prepared to let Bobby off the hook because he was her lover. "My spies tell me that Jacqueline is already blaming Bobby for her troubles . . . As more and more of the deleted passages are leaked to the press, this squabble will increase," presidential adviser John Roche—who considered Bobby "a demonic little shit" and "an arrogant little schmuck"—informed President Johnson,

---

* Although fewer than 5,000 words were finally deleted from the book as a result of the Kennedy intervention, as *Time* (April 7, 1967) pointed out, "Some could have made quite a difference."

who by this time loathed and distrusted Bobby for his own reasons as much as Onassis did for his.[21]

Bobby became so anxious about the public reaction against him, as well as the harm Jackie was doing to her own image—the image that probably mattered even more to Bobby's political aspirations than it did to the Holy Widow's own self-esteem—that he insisted she publicly forfeit the federal money that had been appropriated to maintain her office after the president's assassination.

Again, Jackie turned to Onassis; reassured that he would pick up her expenses, she agreed to the terse, almost testy statement Senator Edward Kennedy's staff wrote for her: "Now that the work at my office, although still considerable, has diminished enough so that I can personally assume the burden of my own official business, I no longer wish a government appropriation for this purpose."

When Gardner Cowles finally rejected Jackie's offer of $1 million to cancel the *Look* serialization, Onassis encouraged her to file suit against Manchester's publishers, Harper & Row. "My God," Bobby's aide Frank Mankiewicz said when he heard what she planned to do, "I think that's a terrible mistake." Bobby, who knew, or must certainly have guessed, whose idea it was, and how little he could do about it, replied bleakly: "Yes, it's a terrible mistake, but nothing can be done about it."[22]

How much Jackie was costing Onassis by the end of 1966 is unclear. Georgakis told me that it amounted to between $80,000 and $100,000 a year; others thought it might have been higher. "Who knows how many envelopes [filled with cash] Jackie got [from Onassis]," Christina Onassis wrote a friend in London in 1987. "Paying for sex have (sic) always been an Onassis weakness," she added, slyly alluding to her own habit of paying for lovers.[23]

But whatever the cost, Jackie was clearly high maintenance and never afraid to ask for more. And although she incensed Onassis with some of her demands, he rationalized her greed as another opportunity to stick it to Bobby—and always paid up.

"If Ari could just have said, 'Fuck you, Bobby' and let it go, his life would have been a lot happier, much simpler—and a whole lot less goddamn expensive. But he couldn't do that," Johnny Meyer would later tell Brian Wells. "He heard Bobby's footsteps behind him all the time."[24]

Whether the footsteps were real or imagined, Onassis's hostility toward Bobby—ignited at a New York cocktail party a decade before—had become pathological by the beginning of 1967, when David Karr began to work his way into Ari's Paris hierarchy.

Now living in France, Karr, Drew Pearson's former legman, had become a considerable wheeler-dealer in his own right. Born David Katz* in Brooklyn in 1918, Karr had undergone a peripatetic and unusual professional ascent: Fuller brush salesman, shipping clerk, reporter for the Communist party newspaper, the *Daily Worker* (where, he later claimed, he had spied on his colleagues for the FBI†). In 1942, after a spell as editor of a weekly newsletter for the American Council Against Nazi Propaganda, he turned up as assistant chief of the Foreign Language Division of the U.S. Office of War Information in New York. A year later, at the age of twenty-five, he met Drew Pearson.

Digging up dirt for Pearson's *Washington Post* column, Karr finally seemed to have found his true *métier*. "David thought nothing of rifling through desk drawers and files in unattended offices," recalled one former associate on the Pearson staff.[25]

It also gave him a taste for blackmail: "Let us say that he knew how to make the most use of what he knew," his former London business partner Ronnie Driver told me. "It might be a little harsh to categorize it as

---

* He never changed his name legally. "I assumed the name Karr about 1937, 1938, for professional purposes as a newspaper man," he told the House of Representatives Special Committee on Un-American Activities in 1943, the year before he became a legman for Pearson.

† In his evidence before the Special Committee on Un-American Activities, Karr claimed that during his association with the *Daily Worker,* he had sent regular reports to the FBI. Later, J. Edgar Hoover's office told the committee's chief investigator that while Karr had visited their offices "once or twice, he was of no service" to them, and they were "not concerned with him one way or the other as an informer. . . ."

blackmail, but that was undoubtedly the area in which he operated. He could destroy people overnight with the things he knew." [26]

Karr most likely first met Onassis in New York in 1956, when Karr was running a public relations company managing proxy fights in corporate takeovers. In 1965 Onassis loaned him $27,000* to buy a movie option on E.L. Doctorow's *Welcome to Hard Times,* which went into production at MGM's Culver City studios the following year. The money was handed over in cash and apparently with little expectation of Onassis ever getting it back—except in favors returned. [27]

And David Karr had already proved that he was in a position to do the kinds of favors Onassis liked best. It was Karr who had planted in Pearson's column the story of Onassis's affair with Lee Radziwill that had upset Bobby so much in 1963. Pearson needed no encouragement to run anti-Bobby stories†; according to political commentator Milton Viorst, he had repeatedly gone "beyond the hard facts" to attack Bobby Kennedy: "Like a metronome, Pearson came back again and again to denounce Bobby in the last year of his life." [28]

Onassis was delighted to have a pipeline to Pearson and regularly called Karr to swap the latest dirt on Bobby. In January 1967, Karr told Onassis a story that changed everything.

---

* This figure varies. Karr told the author that Onassis had "a small investment" in his production company: "perhaps $15,000"; Onassis claimed it was $50,000; Georgakis and Johnny Meyer independently came up with the figure of $27,000.

† Pearson had never forgiven the Kennedys for suing him for alleging that John Kennedy had not been the true author of the Pulitzer Prize–winning *Profiles in Courage.* (Walter Isaacson and Evan Thomas, *The Wise Men,* p. 591. London: Faber, 1986.)

FIFTEEN

TOUT PASSE

Nothing is so burdensome as a secret.

—FRENCH PROVERB

D avid Karr loved to brag about the people and things he knew, and Onassis loved to hear gossip about the high and mighty. A few weeks ago, Karr began in his peremptory way when Onassis picked up the phone, Drew Pearson got a tip from a client of Washington lawyer Edwin Morgan who claimed that shortly after the Bay of Pigs debacle he had been hired by the CIA and Attorney General Robert Kennedy to kill Castro. The source was sufficiently reliable to persuade Pearson to take the story to Chief Justice Earl Warren. But when Warren gave it to the Secret Service, who passed it over to the FBI, fearing that it would disappear into the maw of J. Edgar Hoover's files, Pearson also took it directly to President Lyndon Johnson.

LBJ believed that President Kennedy had been assassinated in reprisal for Bobby Kennedy's attempts to get rid of Castro, and that was the reason why the "grandstanding little runt"[1] did not want his brother's murder investigated*; so he lapped up Pearson's story. The Kennedys seemed

---

* Johnson told television newsman Howard K. Smith, "President Kennedy was trying to get Castro, but Castro got him first." (*New York Times,* June 25, 1976.) But even

to have been "operating a damn Murder Incorporated in the Caribbean," he told Leo Janis of *Time*,[2] as he embarked on a roundelay of insinuation and gossip mongering, slyly lighting a fuse he hoped would "blow Bobby out of the water"[3] before the 1968 presidential election campaign even got started.

Two months later, Karr sent Onassis a couple of newspaper clippings. The first was a Pearson column in which he finally broke the story: "President Johnson is sitting on a political H-bomb: an unconfirmed report that Senator Robert Kennedy may have approved an assassination plot which then possibly backfired against his brother."[4]

The second clipping claimed that Bobby was also implicated in Marilyn Monroe's death. Onassis sent this to his old friend, Monroe's publicist, Rupert Allan.

"It was from one of those insider newsletter publications. The headline was something like *The Mysterious Death of Marilyn Monroe*," Allan told me.* "About twenty minutes after it arrived by courier, Onassis was on the phone: 'Do you think it's true, Rupert?' The honest answer at that time would have been, 'I don't know,' " Allan said, although some believed that he himself had been part of the cover-up, and his own story about what happened on the night Monroe died changed subtly over the years.†

But Allan's dislike of Bobby was unwavering. "Bobby Kennedy was a terrible person, he and his brother had treated Marilyn like shit, so I said,

---

though he had said "President Kennedy," as Jeff Shesol points out in his incisive study of the two men, Johnson knew that it was Bobby Kennedy who had "tried to get Castro." (Mutual Contempt, p. 132.)

* Probably "The Strange Death of Marilyn Monroe," according to Peter Collier and David Horowitz (*The Kennedys: An American Drama,* p. 323. New York: Summit Book, 1984), this was a privately published right-wing pamphlet "charging that Bobby had been having an affair with the film actress and, when she threatened to expose some of his dealings in appeasing the Castro regime, had her killed."

† According to investigative journalist Matthew Smith, Allan attended a "strategy meeting" held in the offices of producer Arthur Jacobs a few hours after Monroe's death at which "plans were laid to contain the situation and provide for the protection of Robert Kennedy." Smith quotes Allan as saying that the cover-up was "carefully done and beautifully executed." (*The Men Who Murdered Marilyn,* London: Bloomsbury, 1996.)

'Yeah, I think so. I think Bobby was involved.' Ari said, 'Yes, that's what I think, too, Rupert.' He was very interested in Marilyn's death. He'd never met her, but he'd offered to lend her his yacht when she was fired, and he seemed to care very much about what had happened to her. But I didn't hear from him again, or find out whether he had found the [Bernard Spindel] tapes [reputedly recorded the night Monroe died] he'd asked me to get for him."[5]

But by this time Onassis had something else on his mind.

Six months earlier, Prince Rainier had created 600,000 nontransferable SBM shares in the name of the principality. In one stroke this cut Onassis's 52 percent domination of the company that owned Monaco to less than one third.

The lesson, Onassis told Johnny Meyer, is: "Never invite royals to your yacht if it's bigger than theirs."[6] It was a nice joke, but Meyer knew how deeply Rainier's ploy angered him. Moreover, although many people believed that Onassis's determination to hold on to Monaco had more to do with pride than good business sense, he needed the principality more than ever: It defined his wealth, and had been his collateral as well as his social power base since the fifties. At his birthday party at the Hotel de Paris in January, he had told Christina and Alexander, now young adults, "I'm sixty-one years old. I'm not going to let some jumped-up princeling steal the pearls from the oysters on my plate."[7]

Age was concerning him a lot. He had actually turned sixty-seven in January, not sixty-one, and he was clearly feeling the strain of a deception that was becoming harder to sustain.

After he had suffered an excruciating pain in his chest one evening on Skorpios, Callas summoned a heart specialist from the American Hospital in Paris. An angiogram showed that his arteries were in good shape for his age; his cholesterol count, however, was high, and an irregular cardiac arrhythmia was detected. Ordered to stop smoking and cut down his drinking, he simply increased his intake of Nembutal.[8]

Meanwhile, the cunning of Rainier's shares offensive had convinced Onassis that the Americans must be behind it; and he was determined not to be finagled out of Monaco as he had been cheated out of Saudi Arabia a decade earlier. That experience felt more like a wound that had never healed than a memory.

When he thought of Jiddah, said Georgakis, Onassis thought of Bobby, whose "blood trade" tirade against Greek shipowners who had dealt with Red China during the Korean war was, he believed, at the root of all his problems with the United States. The accusation was as inevitable as it was unfair, but by this time his suspicion had degenerated into a phobia about Bobby's power and persecution of him.

Complicating matters still further was the fact that although he claimed that Jackie had agreed to marry him, she continued to prevaricate. The problem, Jackie said, was how to break the news to the Kennedy family—by which she meant Bobby, of course. She had withdrawn from close contact with the rest of the family, only visiting Palm Beach for Christmas, and Hyannis Port briefly in the summer; she did not want her children to grow up in thrall to the Kennedy ethos, she told friends.[9] "Jackie was still seeing other men, fellows like David Ormsby-Gore [Lord Harlech, the former British ambassador to the United States, and an old friend of the Kennedys] were considered to be her real suitors, but there was never any question [after the 1963 cruise] that she would eventually marry Ari," his sister Artemis Garofalides told me.

Nevertheless, Jackie was pressing Ari to clean up his act; with a marriage around the corner, the last thing she wanted was for him to get into another big confrontation with Washington. Yet he knew that if the Americans were backing Rainier, a fight was inescapable. He knew that the share creation ploy had plunged Rainier's treasury into the red, and his instinct was to hit him with an expensive proxy counteroffensive— which David Karr, an expert in proxy fights, was eager to manage. "It will mean blood on the palace walls. It won't be for the squeamish. It'll be the kind of brawl Ari loves," predicted one French banker.[10]

Still being urged by Jackie to put his trust in the law, Onassis continued to hesitate. It was at this point that he got a call from a figure from the past: Baron George de Mohrenschildt, with whom he had nearly gone into business five years earlier in Haiti.* A petroleum engineer by profession, de Mohrenschildt was now teaching French at a small black college in Dallas.

The reappearance of an old friend down on his luck should not have been a problem for Onassis: A discreet handout, a sinecure somewhere in one of his companies, could have settled the matter quickly. But de Mohrenschildt had reappeared as a darker, more complex character than the dashing playboy-adventurer Onassis had first met at his lawyer Tito Arias's home in Panama City in 1961.

Onassis had been used to meeting shady people at Tito's parties; his lawyer's first love was politics, and politics was even more corrupt in Panama than in most places. Tito's wife, the English prima ballerina, Margot Fonteyn, who would herself be jailed and then deported for aiding one of his plots for a *coup d'etat,* enjoyed the company of the arms dealers, spies, and other dubious characters who mingled at their gatherings on smart Avenue Balboa in Panama City. It was even rumored that Tito liked to share her with other men, and Onassis had imagined that de Mohrenschildt—"an exciting man to look at," according to one woman who knew him[11]—had been invited to the party for Margot's pleasure.

"Tito liked to watch," claimed Onassis, who had seen most sides and combinations of human nature aboard his yacht.[12] And whether this was true or not, the couple plainly liked to shock. Anthony Montague Browne, Sir Winston Churchill's private secretary, recalls an occasion when he was passing their stateroom on the *Christina,* and Tito "threw open the

---

* Onassis claimed that he had backed out of the deal when he saw the terms of the contract de Mohrenschildt had signed with President "Papa Doc" Duvalier. "If we had found oil, we wouldn't even have had an explorer's cut," he said. In his testimony to the Warren Commission, de Mohrenschildt said that on March 13, 1963, he concluded a contract with the Haitian government which guaranteed that he would be paid $285,000 for a geological survey of Haiti to plot out oil and geological resources on the island: $20,000 was paid in cash and the remainder was to be paid out in a ten-year concession on a sisal plantation. (Staff Report of the Selection Committee on Assassinations, U.S. House of Representatives, Ninety-fifth Congress, Second Session, p. 55. March 1979.)

door, and lying on the bed was the lissome and totally nude Margot. Neither was disconcerted. Margot giggled and Tito said, 'Anthony, don't you think that our girl sometimes looks like wet seaweed?' "[13]

But whatever other purpose he had at the Arias party on that evening in 1961, de Mohrenschildt asked the ballerina to appear at a gala charity show he was organizing to raise money for a cystic fibrosis charity. He told Fonteyn that Jacqueline Kennedy was a friend of his, and also the honorary chairman of the National Foundation for Cystic Fibrosis, which he had founded after the death of his son from the disease in 1960. "I said if the dates worked out, I'd be happy to appear," Fonteyn told the author. "But I never heard from him again."

Later, de Mohrenschildt's friendship with Jackie Kennedy caused a furor when it was discovered that he had also been Lee Harvey Oswald's closest friend in Dallas in the months preceding John Kennedy's assassination. Although the Warren Commission "found no evidence linking de Mohrenschildt in any way with the assassination,"[14] and despite the strong possibility that he had been watching Oswald on behalf of the CIA, the mud had stuck.

Jackie had made it clear to him that she never wanted to see him again,* and this was why Onassis was suspicious when de Mohrenschildt called him in the spring of 1967 and said he had something of great importance to tell him; de Mohrenschildt refused to be more specific on the telephone and asked Onassis to send him an airline ticket to Paris.

---

* At the end of his testimony to the Warren Commission, de Mohrenschildt had received an extraordinary invitation from, as he put it in the book he was writing at the time of his death in 1977, "Jacqueline Kennedy's mother and her stepfather, Mr. Hugh Auchincloss," to dine at their home in Georgetown. Apart from the Auchinclosses and de Mohrenschildt's wife, Jeanne, the only other known guest was the former CIA chief Allen Dulles. They talked about the assassination; at one point, Janet Auchincloss wept and embraced Jeanne de Mohrenschildt; later Dulles asked him "a few astute questions about Lee (Harvey Oswald)." But as he was leaving that evening, Janet dropped her hostess's charm and told him coldly: "Incidentally, my daughter Jacqueline never wants to see you again because you were close to her husband's assassin." (George de Mohrenschildt, "I Am a Patsy!", pp. 225–228; HSCA Vol. XII, Appendix.)

Understandably, Onassis did not want de Mohrenschildt within ten thousand miles of him.

But suspecting that the Americans were backing Rainier—and aware that de Mohrenschildt had good connections with the CIA[15]—Onassis was also anxious to find out what he knew. Since he also owed de Mohrenschildt a favor (for not dragging Onassis's name into his testimony to the Warren Commission*) and guessed that de Mohrenschildt would sooner or later come to the same conclusion, Onassis dispensed Johnny Meyer to Dallas with a bagful of dollars to find out exactly what de Mohrenschildt knew.[16]

The story Meyer brought back was more disturbing than Onassis could possibly have imagined: One of the key men behind Rainier's campaign to regain Monaco was Robert Aime Maheu, who, a decade earlier, had organized the CIA plot that destroyed Onassis's deal with the Saudis. Maheu's handling of that operation had made him a legend in the Agency, and the first choice to run a CIA plot to kill Castro[17]—which de Mohrenschildt obligingly traced back through layers of obfuscation and cover-up, through CIA agents, Mafia dons, and Cuban exiles, to the desk of then-attorney general Bobby Kennedy.†

---

* This was a discretion de Mohrenschildt would not repeat in his unfinished autobiography, "I Am a Patsy!" "If you, dear reader, are interested not in the assassinations but in organized murder for profit," he would begin a strangely digressive reflection on how Onassis had made and consolidated his fortune. "Some will say that the introduction of the late Aristotle Onassis in these chapters may be in bad taste—others may find an interesting and significant relevance . . . If you believe in just punishment, Aristotle's rotten soul will remain forever in the Greek-Orthodox hell," he concluded cryptically, shortly before blasting himself through the mouth with a 20-gauge shotgun in Palm Beach in 1977. (George de Mohrenschildt, "I Am a Patsy!", pp. 225–228; HSCA Vol. XII, Appendix.)

† " 'It's got to be set up so that Uncle Sam isn't involved—ever,' " Maheu would later recall his own precise instructions to Las Vegas Mafia figure Johnny Rosselli to arrange Castro's murder for the Agency. " 'If anyone connects you with the US government, I will deny it,' I told him. 'If you say Bob Maheu brought you into this, that I was your contact man, I'll say you're off your rocker, you're lying, you're trying to save your hide. I'll swear by everything holy that I don't know what in hell you're talking about.' " (Robert Maheu and Richard Hack, *Next to Hughes,* p. 139. New York: HarperPaperback, 1993.)

De Mohrenschildt could have told Meyer no better story to exacerbate Onassis's paranoia, of course.* And to add to Onassis's misery, and his rage, de Mohrenschildt had also claimed that Stavros Niarchos—who had remained close to the CIA after his help in the destruction of Onassis's Saudi deal—was also part of the team backing Rainier's bid to run Onassis out of town.† At least one of his companies (Niarchos London, Ltd), was a CIA proprietary,‡ fronted by former Athens station chief, Al Ulmer.

Onassis was pole-axed. Clearly, it was all connected: the CIA, Niarchos, Maheu, Bobby Kennedy, a labyrinth of enemies, interests, agencies, inextricably linked in a conspiracy to destroy him. How could it not be? To Onassis's increasingly paranoiac mind it was all too clear.

---

* When I repeated de Mohrenschildt's story about his part in the plot to get Onassis out of Monaco to Robert Maheu in Las Vegas in 1995 he denied it completely. Looking like the kind of Vatican banker who kept an expensive mistress on the Via Sistina, he said pleasantly: "I wasn't in Monaco in 1967. I wasn't involved in that business." I told him that Johnny Meyer claimed that he had seen him driving away from the Hotel Hermitage, accompanied by Stavros Niarchos's English public relations consultant, Alan Campbell-Johnson (a claim that the Englishman did not dispute but declined to explain when I asked him about it in London in 1997). "Johnny Meyer saw somebody else . . . people often think they see me where I am not. I guess it's part of my mystique," Maheu smiled cherubically, as if he didn't expect to be believed, but was too rich, and too old, to let it worry him anymore.

† But even the well-informed de Mohrenschildt did not appear to know that Ulmer and Maheu had worked together in the past: Ulmer had been the CIA's Far East Division chief with the job of overthrowing Indonesia's president Sukarno, a high-rolling nationalist leader who was playing off the Americans against the Russians, when Maheu made his notorious CIA pornographic movie with a Sukarno lookalike frolicking with a blonde in a Kremlin bedroom to discredit the Indonesian leader. (Assassination Report, p. 74n; Burkholder Smith, *Portrait of a Cold Warrior,* pp. 240–242. New York: Putnam's Sons, 1976.)

‡ CIA proprietary: "These are ostensibly private institutions and businesses which are in fact financed and controlled by the CIA. From behind their commercial and some times non-profit covers, the agency is able to carry out a multitude of clandestine activities—usually covert-action operations." (Victor Marchetti and John D. Marks, *The CIA and the Cult of Intelligence,* p. 134. London: Jonathan Cape, 1974.)

But nobody will ever know what game de Mohrenschildt was playing, or for whom. In Norman Mailer's words, he possessed "an eclecticism that made him delight in presenting himself as right-wing, left-wing, a moralist, an immoralist, an aristocrat, a nihilist, a snob, an atheist, a Republican, a Kennedy lover, a desegregationist, an intimate of oil tycoons, a bohemian, and a socialite, plus a quondam Nazi apologist once a year."[18] But whichever side de Mohrenschildt was on this time, his intervention was decisive. "It knocked the wind out of Ari's sails completely . . . all those forces aligned against him, Ari felt he didn't have a chance," said Georgakis, who believed Ari would have fought much harder for Monaco if de Mohrenschildt hadn't intervened.[19]

Meanwhile, Onassis's foreboding deepened as his petition to stop Rainier's share creation ploy moved slowly through the Monaco supreme court. In March, the court found in favor of the prince. "I played by the rules and lost. I'll never make that mistake again," Onassis told his old friend Rico Zermeno.[20]

For more than a decade he had been the most powerful man in the principality, and now that he was just another rich punter, he sold his holding back to Rainier for $9.5 million. On his last night in Monte Carlo, Onassis dined on the terrace of his beloved Hotel de Paris with his children Alexander and Christina, and a few old friends. He knew he was part of the show as he sipped vintage Taittinger Comtes de Champagne, his beautiful yacht waiting in the harbor below to bear him off like a departing king.

It was a wonderful performance. It hurt him deeply to know that never again would he be able to regard the principality in the way he had when it belonged to him. Jiddah had been his biggest business failure, but losing Monaco hurt him far more; for it tore at the heart of his self-esteem.

"Everybody knew that he had suffered a humiliating defeat, but he wanted the world to know that he believed that nothing in life is worth worrying about," remembered Georgakis. "He said, 'Yannis, ten years ago it would have mattered, but now—I will always remember the way he raised his glass— 'Tout passe.'"[21]

SIXTEEN

BREATHING NEW LIFE INTO OLD RUMORS

**If I am pressed to say why I loved him,
I feel it can only be explained by replying
"Because it was he; because it was me."**

**—MICHEL EYQUEM DE MONTAIGNE,
1533–1592 (ESSAY 1580)**

Now that the "Monaco problem" had been "resolved" (nobody around Onassis mentioned a fight, nobody spoke of defeat), Callas prayed that their own difficulties could also be put behind them. She had been married to Onassis in all but name for five years. She loved the high life he had given her, and she would miss Monte Carlo almost as much as he would, but his failure had made him even more dear to her.

But the affair for which she had sacrificed her career, her marriage, and her pride was already over for Onassis. Callas had opened doors for him and given him an aura of class, but by 1967, she had served her purpose. With Jackie waiting in the wings, the affair would have ended with his exit from Monaco—had they not been bound together in an action in an English court of law.

*   *   *

Three years earlier, Callas had acquired a 25 percent stake in a bulk carrier called *Artemision II*. Her old friend Panaghis Vergottis also owned 25 percent, and Onassis 50 percent, out of which he made a present of 26 percent to Callas, giving her a controlling interest in a vessel to be managed by Vergottis. But on her maiden voyage the tanker developed engine problems. Losing her nerve, Callas accepted Vergottis's offer to convert her investment into a straight loan. This was on the understanding, she believed, that if she changed her mind (i.e., if the vessel became profitable) she could buy back in. But shortly after this, she fell out with Vergottis, and when she attempted to revert to the original deal, he refuted her claim completely.

In the ironic nature of human relationships, Jackie this time had begged Onassis *not* to go to law. It took no great prescience to see that a court case in London would turn into a media circus, of course, and it duly lived up to Jackie's worst nightmare.

Dressed in a scarlet gown and white twenties-style turban, her face heavily made up, Callas arrived on the first day clinging to Onassis's arm as if they were attending a fashionable first night. With allegations of fraud, treachery, and criminality in the air, and with heavy intimations of sex and jealousy at the root of the action, Vergottis's counsel was keen to establish the intimate and sensational nature of the relationship between Onassis and Callas.

There were many exchanges like this one:

"After you got to know Madame Callas did you part from your wife and did Madame Callas part from her husband?" "Yes, sir. Nothing to do with our meeting. Just coincidence." "Do you regard her as being in a position equivalent to being your wife, if she was free?" "No. If that were the case I have no problem marrying her, neither has she any problem marrying me." "Do you feel any obligation towards her other than those of mere friendship?" "None whatsoever."

It could have been scant consolation to Jackie, who was furious at the way she felt Callas had maneuvered Onassis into the role of co-litigator, and even less comfort to Callas who, although she understood the expediency of his answers, knew that his denial of all obligations toward her went beyond legal necessity: It wiped out the past; it was a public declaration of intent.

Vergottis had known Onassis since the thirties—and might even have been his lover.* "I have heard so many things [about Onassis] that it would make your hair stand on end," he told the judge. "The things he has done and how he started. I have been to Greece and investigated lots of things. *He is black in his heart.*" [1]

Outbursts like this breathed new life into old rumors about Onassis's past, which Jackie wanted so much to be forgotten. Every evening she would call from New York, telling him how the story was playing in the States, and how painful it was for her to read.

Early on the morning of Friday, April 21, the fourth day of the case, the telephone rang in his suite at Claridges, waking him from a deep sleep. It was Georgakis calling from Athens to tell him that the colonels had staged a *coup d'etat.*

They both knew that a military coup was in the cards to preempt what threatened to be a resounding victory for George Papandreou's liberal Centre Union party at the May elections. King Constantine had thrown out a Papandreou government in 1965, and a cabal of monarchist generals was determined to thwart his return. The great weakness of their coup, however, had been the large number of people who knew about it. And now the colonels had beaten their own generals to the punch.

Onassis had no ideological agenda of his own, although he usually had at least two or three members of the cabinet in his pocket no matter which government was in office. (*Bribery* is not a word diplomats like to use, and the British Foreign Office referred only to his "considerable degree of political influence" and his "economic interests" when they raised the possibility of recruiting Onassis as an agent in 1964, after an

---

* Callas had hinted at this to friends, and it was the reason why one of her English lawyers would later tell the author that "It was a sort of catharsis, a sex case in the end."

intelligence report revealed his closeness to prime minister Papandreou.*)

But the colonels' coup had shuffled all the cards. Onassis knew that he would have to tread carefully until he found out where the colonels stood in relation to Washington and what role the king would play in their plans.† For although the *junta* would offer plenty of possibilities for a man like Onassis, the king was a Stavros man, and Stavros was a CIA asset, and the next few months would be extremely difficult for him.

---

* A senior NATO official on his way to Athens for urgent talks with the government about the growing belligerence between Greece and Turkey was disturbed to learn that he had been given "only half an hour with the foreign minister and a quarter of an hour with the prime minister right at the end of his two-day visit." When he expressed his anxieties to Onassis, a fellow passenger on the plane to Athens, Onassis told him to "ignore the official programme . . . and go to his hotel and wait for a quarter of an hour. After ten minutes Mr. Onassis rang him up to say that they were dining with the prime minister and the foreign minister that evening; and at this dinner Mr. Onassis violently, obstinately attacked Mr. Papandreou for incompetence, obstinacy, etc., etc. They even called in young [government minister Andreas] Papandreou who was given a dressing down by Onassis in similar terms." Although he agreed that "even at the highest level, [Onassis] can fix things at short notice," the Embassy's first secretary Richard Sykes (who was almost certainly also an MI6 agent, and would later be murdered by the IRA) responded that he thought "there might be dangers in trying to do anything *openly* in this way." (Confidential Foreign Office memorandum to British Embassy, Athens. July 8, 1964; Foreign Office document July 16, 1964. Public Records Office, London: FO371/174 IC4389.)

But did the British use Onassis covertly? "Onassis was always regarded by us as the Brits' Greek, just as Niarchos was our Greek," CIA man Miles Copeland told me shortly after the publication of *Ari* in 1986.

† Two of the *junta* leaders—Colonel George Papadopoulos and Colonel Nicholaos Makarezos—held key posts in the KYP, the Greek intelligence service, which had close ties with the American agency.

SEVENTEEN

AN OLDER WOMAN

**There is no such thing as an older woman.
Any woman of any age, if she loves, if
she is good, gives a man a sense of the
infinite.**

—JULES MICHELET, 1798–1874

I t was in this hiatus that Onassis turned his attention to a family
matter that had been on his mind for some time, and about which
his former wife Tina, now the marchioness of Blandford,* was
becoming increasingly hysterical. Although they had seldom agreed on
anything, on this they were unanimous: Their son, Alexander, nineteen
years old, heir to two great shipping fortunes, was the catch of the year;
and Fiona Thyssen, aged thirty-five, a divorced mother of two young
children, was ruining his life.

There was no mystery about Alexander's fascination for women
who were older than himself. At fourteen, one of his father's male shipping assistants, who took care of Alexander during his parents' frequent

---

* She had married the marquess of Blandford, son of the duke of Marlborough and a
kinsman of Winston Churchill, in 1961.

absences, introduced him to his own women friends, most of whom were in their thirties. Jacinto Rosa, Onassis's chauffeur, who had known Alexander since he was twelve years old, and had taught him to drive, was disturbed by the younger Onassis's precocious interest in sex. "Many times he asked me to drive him to the Bois de Boulogne, where he liked to spy on the prostitutes working with clients in their cars," Rosa said.[1]

Alexander was sixteen when he started work in his father's Monaco office, having flunked his exams after returning several days late from a tryst in the South of France with Odile Rodin, the widow of playboy Porfirio Rubirosa. Although Onassis took a keen interest in his son's sexual adventures ("what's bred in the bone comes out in the flesh"), he refused to "piddle away good money on a lazy kid."[2] Alexander suspected that the truth was "the old man didn't want me to be better educated than he was."[3]

Nevertheless, Alexander was not without accomplishments—albeit, perhaps, those of any young man waiting to inherit a fortune: He spoke excellent French, colloquially perfect English, with a Parisian accent, and some Italian (his Greek, like his sister's, was not so great); he was a fine tennis player, an expert skier, played backgammon like a pro, drove fast cars with skill, and was learning to fly his own plane.

Later he told me that not a day had passed when he had "not been intimidated by the old man's wealth."[4] And although he recognized the miracle his father had performed in creating his extraordinary fortune, his discontentment was turning to hostility by the time he fell in love with the Baroness Thyssen-Bornemisza.

At first, Onassis dismissed the affair as an adventure that would pass in the night: Women like the Baroness are a necessary education, he told friends.

She was a complex and beautiful woman. As Fiona Campbell-Walter, she had achieved fame as an international fashion model, and was still only twenty-three when she retired in 1956 to marry the multimillionaire German industrialist and art collector Baron Heinrich Thyssen-Bornemisza. Alexander had first seen her when he was twelve years old. Climbing out of a sports car in St. Moritz in a snowstorm, wearing a full-length black leather coat, her long red hair tumbling from beneath a black chinchilla hood, she had skin that old orchid horticulturists must dream about. Alexander thought she was the most exciting woman he had ever seen.

Six years later, when invited to a dinner party by his mother, he said he would go only if Fiona Thyssen was invited, too. That night, to Thyssen's surprise, and with no small sense of guilt—"being sixteen years older, I didn't know whether he saw me as a substitute mother figure or what"—they went to bed together. Thyssen treated it as a one-night stand. Alexander wanted a *commitment*.[5]

Divorced from one of the richest men in the world, still a young woman, with money and a social position, Thyssen wanted to be married again—but not to Alexander. He was too young for her, she was too independent for him; anyway, she knew enough about the politics of Greek dynasties to know that his bride would have to be a rich virgin from an approved shipping lineage.

She did her best to break free. He refused to let her go. Soon it was very clear to both of them that they had embarked on a very dangerous affair. Eventually, they accepted that "we had become indispensable to each other and should just try to survive together on a daily basis, and whether it lasted a week or a month or a year we should enjoy it and be grateful," Thyssen said.[6]

"I think my mother was one of the last to know about their affair," Christina Onassis told me. "She was simply furious with Fiona for seducing her son. 'He is still absolutely a child,' she said, which, of course, was not true at all. My brother had been sleeping with women since he was fourteen years old."[7]

Fiona Thyssen was three years younger than Tina Blandford. They went to the same dinner parties, attended the same balls, and danced with the same rich men. "Tina didn't give a damn about whether Fiona was good for Alexander or not," said a mutual friend. "She was thinking of herself, and the prospect of Fiona becoming her daughter-in-law was too unbearable to contemplate! She went to extraordinary lengths to stop it."

Thyssen had been a call girl, Tina claimed; she knew a man who had paid her fifty pounds to go to bed with him when she was seventeen years old; now her fee was a Patek-Philippe watch, Tina claimed, although king Farouk had once paid $100,000 for one night with her at Badrutt's Palace Hotel in St. Moritz. . . . Tina's stories became increasingly lubricious as they became more libelous.* Alexander and Thyssen played a game in which they tried to guess her next calumny. "We were never even close. She always surprised us," Thyssen told me.[8]

Tina began pressing Onassis to deal with the problem. She knew that he was deeply jealous of Alexander's success with Thyssen. She was the kind of trophy woman Onassis himself might have expected to seduce, and he resented the relationship every bit as much as Tina did. But with so much else on his mind, he had been hoping that the affair would burn itself out.

The loss of Monaco, the increasing financial difficulties with his airline, the unresolved court case in London (he and Callas had won the first round, but a new trial had been ordered on appeal), the complications of his private life—even perhaps a psychic toll of conscience over what he was about to do to Callas—were all causing him stress. "Jackie, Maria, and Tina all wanted something from him," said Georgakis. "He would put the phone down on Maria, and Jackie would be on the line. He'd deal with Jackie, Tina would call. It made us smile, but it was getting to him . . . he was on a very short fuse."[9]

"He's Greek. When he loses his temper all hell breaks loose," Tina had once warned Nigel Neilson, a London public relations man hired to burnish Onassis's image.[10] Tina knew, therefore, that she was stirring an already brewing crisis of explosive dimensions for her son and his lover.

But was there another reason why Tina was making such a fuss about their son's affair? Was it, as Christina Onassis would later suspect, to

---

* Partly because of Tina Blandford's stories, Thyssen was rumored to be the beautiful model—with hair the pale red of a winter sunset and eyes as green as emeralds, married to the richest man in Germany, perhaps in all Europe—for Kate McCloud, Truman Capote's heroine in his jet set *roman à clef Answered Prayers* (London: Hamish Hamilton, 1986). Other suspects included Lee Radziwill and Pamela Churchill Harriman.

divert Onassis's attention from the increasingly serious affair she was having with her brother-in-law, Stavros Niarchos?

In December 1967, the Greek king's counter-coup collapsed almost before it began, and he fled with his family to Rome. (Onassis cabled Niarchos: *"I SEE YOUR EXALTED FRIEND NOW PLAIN MR. COSTA P. GLUCKS-BURG, COMMONER OF NO FIXED ABODE."* [11]) Washington, which had expressed a cautious recognition of the *junta*—provided that it remained loyal to NATO, refrained from maltreating political prisoners, and moved toward democracy—now embraced it.

Colonel Papadopoulos and his fellow conspirators were not stupid men, but "their political ideas were crude and naive," [12] and they were putty in Onassis's hands. Through a nexus of favors and flattery, many of them soon found themselves deeply indebted to him: a girl from Madame Claude's for this colonel, a handsome youth for another, and beluga for the generals of the future.

But Papadopoulos got the best of it: cruises on the *Christina;* the permanent loan of Onassis's villa in exclusive Lagonissi; and when he ordered forty *haute couture* dresses at one thousand dollars each for his wife, Onassis picked up the tab. [13] "Colonel," Onassis told him by his own account at the outset, "we're both people users, so let's do what we do best and see what happens." [14]

As their friendship blossomed, Onassis's plans became more ambitious. Under the code name Omega, he began putting together what he would later call "the biggest deal in the history of Greece." * He asked David Karr to sound out the Soviets about the possibility of supplying crude oil for a refinery he planned to build near Athens.†

---

* This eventually became a four hundred million dollar (four billion in present-day dollars) package of investments, including the construction of an oil and alumina refinery, an aluminum smelter, a power station, shipyards, and an air terminal.

† Although on another occasion Karr would credit Armand Hammer with helping him establish his first Soviet links some three years after this date, his London partner Ronnie Driver says that "David probably had some valid and very good reasons" why he did not want to admit his earlier excursions to the Soviet Union for Onassis.

"I said, 'I imagine the Americans might have a thing or two to say about that, Ari,' " Sir John Russell, the former British diplomat, who was now chairman of Elf Oil, remembered of his reaction when Onassis told him about Omega. "He was talking about acquiring the rights to handle the oil from beginning to end: refining it in his own refinery, transporting it in his own tankers, selling it through his own distribution network, Lord knows what else besides." [15]

Even though he knew that what Onassis said was often exaggerated and never quite accurate ("One was always struggling after the truth with Ari," he claimed) the Englishman was astonished at the sheer audacity of the deal. "It was Jiddah all over again in many ways. I said, 'It's going to flutter the dovecotes in Washington, isn't it, Ari?' He then made this interesting remark: He said, 'A marriage of interests can solve many problems, my dear fellow.' I missed its significance at the time; it only struck a chord when he married Jackie Kennedy." [16]

Almost everything Onassis did in life had its roots in some earlier experience. In the forties, he had outflanked the New York Greek shipping establishment, which had been hostile to him and his ideas, by marrying one of its princesses—the powerful Stavros Livanos's daughter, Tina.* And although it was a stretch to think he could checkmate Washington with the same ploy, he believed that the Americans would be less keen to interfere when he involved the Soviets in Omega if he were married to their own princess.

---

* In 1946, when Congress made U.S. Liberty ships available to Allied operators, the Greek government had authorized the Union of Greek Shipowners in New York to act as its agent. On offer at $550,000—$125,000 down, the balance payable over seven years at an interest rate of 3 percent—the ships were a steal. But dominated by those rich and powerful owners who had moved to New York at the outbreak of war, the union decided that one hundred ships would be sufficient for their needs. However, when Onassis asked for thirteen, he was told that none was available. His future father-in-law Stavros Livanos had gotten the dozen he wanted, and so did the rest of the senior New York Greeks. Onassis, they had made it perfectly clear, was not one of them, and had never even been in the running. Ironically, one of Livanos's wedding gifts to him was a Liberty ship—with a four hundred thousand dollar mortgage still attached.

*   *   *

Exactly when Callas became aware of what was happening is unclear. But her determination to exhibit grace under pressure gave her a poignancy that was rare in her stormy relationship with Onassis. A friend recalled overhearing one sad exchange on Skorpios late in the summer of 1967: "Maria said, 'With a woman like her,' she meant Mrs. Kennedy, 'and a man like you . . . starting something is easy, Aristo. But how do you stop it?' "[17]

The evening before Callas was to return to Paris at the end of that summer—and Onassis to Athens and the colonels with whom he had mixed—they talked as if they both knew they had spent their last summer together. Onassis said, "The only free people are those who love nobody." She thought that was "too big a price to pay" for freedom. He asked her what she most wanted for herself, one of the half-dozen guests at dinner that night later recalled. "I just want to be on good terms with myself," she answered.

During the following months they saw little of each other. She went for weeks without leaving her apartment on avenue Georges-Mandel; she told friends that she was pleased to have climbed off Onassis's merry-go-round; that she did not miss the impromptu dinner parties, the all-night drinking bouts in nightclubs; she did not miss the fights.

But the wistfulness when she talked of the past was never quite concealed.

EIGHTEEN

A TERRORIST BY ANY OTHER NAME

**Only a Greek should share the secrets
that lie in a Greek's heart.**

—CONSTANTINE GRATSOS, 1905–1981

One evening in January 1968, David Karr and Johnny Meyer met for drinks at the Plaza-Athenée in Paris. Karr was accompanied by a man he introduced as Dr. Michel Hassner, an investment consultant with the Arab Bank. Later they were joined by Paul Bougenaux, the hotel's concierge, whose saturnine good looks and dark business suits, Meyer later remarked, gave him the "air of a society abortionist."[1]

At first, Meyer imagined that Hassner was there because he was involved in a property deal he knew Bougenaux and Karr were putting together (they were plotting to acquire the Plaza-Athenée, Georges V, and La Trémoille hotels for the British hotel magnate Sir Charles—now Lord—Forte*). However, after Bougenaux left, Karr turned to

---

* The problem Karr and Bougenaux had to overcome was this: President de Gaulle had proclaimed the three hotels to be part of France's patrimony, and said they must never

Meyer. "Johnny, Dr. Hassner here is the man who can save Ari's ass," he said.[2]

Not Mr. Onassis's ass exactly, Hassner smiled. But he might be able to help save his airline. He specialized, he said, in aviation finance and had some ideas about how Onassis might restructure the debt and at the same time expand his Olympic fleet. Could Meyer arrange an appointment with Mr. Onassis for him? It was no secret that Olympic Airways was in trouble and needed regular infusions of operating capital. It did not require a great deal of imagination to spot that Karr was not there simply as a go-between and that Meyer was wary of being drawn into one of his intrigues. Meyer knew exactly why Karr did not want to arrange the meeting himself: If the Greeks around Onassis suspected that Hassner was Karr's man, they would oppose everything he suggested.

Gratsos did not trust Karr. "Fuck does he know about oil, Ari?" he had complained when Onassis told him that he planned to ask Karr to sound out the Russians about supplying Soviet crude for the Omega refinery. "David knows about *men*," Onassis said.[3] Karr had accomplished something extraordinary in getting so close to Onassis; no other non-Greek had ever penetrated the organization so deeply. "Only a Greek should share the secrets that lie in a Greek's heart," Gratsos said.[4] This profound thought was shared by Georgakis—but probably because he had "a strong sense that Karr had his eye on his job at Olympic Airways," said the mischievous Nigel Neilson, Onassis's public relations man in London.[5]

But whatever misgivings Meyer might have had about becoming a stalking horse for Karr, he was proud of his access to Onassis. And flat-

---

be permitted to fall into foreign hands. And so, circumscribed by patriotism on one hand, and beset by serious union problems on the other, their owner Mme. François Dupré was stuck with what was rapidly becoming a trio of white elephants. However, as well as senior concierge, Bougenaux was also the group's chief union organizer. This was a source of considerable power, and after he had made union objections to a foreign sale quietly disappear, along with the restrictive practices that had discouraged other potential buyers, Karr took the deal to Forte. Shortly after Forte got the hotels, Bougenaux was made general manager of the Paris group. David Karr was given a substantial finder's fee, plus a $100,000-a-year directorship—"a little *douceur*," he called it with his growing fondness for the genteel obfuscation of the French language.

tered, if not fooled, by Karr's assertion that he was the only man who could get Dr. Hassner into avenue Foch, he agreed to arrange a meeting when Onassis returned from Athens.

Dr. Michel Hassner's real name was Mahmoud Hamshari, and had history been kinder he would probably have lived and died an orange farmer in the village of Um Khaled, near Jaffa, where he was born in 1939. Instead he became a member of the Palestinian guerrilla group Fatah. In June 1967, following the humiliating defeat of the Six-Day War—when Israel seized all of the Sinai Peninsula, including East Jerusalem, the very heartbeat of Palestinian nationalism—Fatah was in disarray, its leadership split. Its military commander, Yasser Arafat, wanted to mount an immediate counteroffensive, with commando raids and an escalation of guerrilla activity in the occupied territories. His critics within the organization, however, felt that his hawkishness ("He wanted to go after the Jews like gangbusters," said one[6]) would simply provoke massive Israeli retaliation when they were at their weakest and most demoralized.

The conservative faction, led by Khaled Hassan, increased the pressure to remove Arafat. Although Hassan's lack of enthusiasm for a swift retaliative strike was dismissed by Arafat's people—known as "the mad ones"[7]—as a lack of fighting spirit, there is no evidence for this, and a growing number of rank-and-file members were coming around to Hassan's more practical viewpoint. Abu Iyad,* Fatah's intelligence chief and an Arafat man, was worried that they might be losing the argument and in danger of splitting the organization irrevocably, so he called a secret meeting between the leaders of the factions to try to resolve the impasse.

In addition to Abu Iyad, those at the meeting, held in a room on the third floor of a hotel in an eastern suburb of Damascus, included Arafat; his deputy military commander, Abu Jihad†; Ali Hassan Salemeh, an Arafat confidant; Kamal Adwan, a key Fatah figure, who later became a

---

* Nom de guerre of Salah Khalef.

† Khalil al-Wazir.

founding member of the Black September organization; and Mahmoud
Hamshari. These, according to one attendee, were "the cream of the mad
ones' mad ones." Opposing them were Khaled Hassan, the moderate
pragmatist, who spoke for the majority of those opposed to Arafat's auto-
cratic style of leadership, and several of his supporters, including Issam
Sartawi, a former heart surgeon, educated in the United States; and
Mohammed Ibrahim, a senior Fatah intelligence agent.*

Although Hamshari was the most junior person on the Arafat side,
there was about him a certain single-mindedness, and when he spoke, he
did not directly support or attempt to defend Arafat's hardline policy
against the Israelis, but began a diatribe against America. With facts and
figures, he condemned the U.S. Pentagon for its operational assistance to
the Israelis in the Six Day War; he attacked Washington's diplomatic support for Israel at the United Nations; he poured scorn on American
politicians and others who spoke up for Israel or sent it money. America
was as much their enemy as Israel was. Yes, but what could they do about
it? he was asked.

It was at this point that Hamshari proposed that they "kill a high-
profile American on American soil"—a relatively simple matter in such
an un-security-conscious country like the United States, he said—in
order to make Washington "think twice about backing the Jews." [8]

The suggestion seemed so extreme that the Hassan moderates suspected it was an attempt to unnerve them, or to lead them into some
sort of trap. Assassinations to remove political enemies, or people who
were giving you trouble, were not unknown, of course, but to contemplate killing at random some famous American in America itself *pour
encourager les autres* was madness.

Surprisingly, perhaps, even the "mad ones" did not seem to think
much of the idea either, and Hamshari moved on to his second proposal.
This was to extend their fund-raising activities to the United States itself.
Washington and the Jewish lobby were their real enemies, he repeated in
a more conciliatory tone, not the American people. "The irony of get-

* Mohammed Ibrahim is a pseudonym for a former Fatah member, whose personal
details I have altered to protect his identity at his request.

ting Americans to help us fight the people other Americans were paying to fight us, I think that perfectly epitomized Hamshari's clever mind," a Fatah source told me.[9]

Under what name and on what passport Hamshari began visiting the United States and France in the late summer or early fall of 1967 is a secret that probably died with Abu Jihad, Fatah's specialist in providing false passports and identities.*

Lord Forte recalls meeting Hamshari—who had metamorphosed into Dr. Hassner by this time—with David Karr and Paul Bougenaux at the Plaza-Athenée shortly after Karr approached him about acquiring the Paris hotels. "He was there on several occasions, always in the background. I don't know what he did. There was always something he didn't want you to know about when you were dealing with David Karr," Lord Forte told me. But Hassner must have had a purpose, he said, because "David didn't have people around who were not of use to him."[10]

Hamshari was in his element in the Paris of fine hotels, expensive restaurants, and attractive, available women. For someone who had been living with terrorists and among refugees for most of his adult life, it was an exhilarating experience, and he seems to have adapted with remarkable ease.

Some ten days after they met at the Plaza-Athenée, Johnny Meyer invited "Dr. Hassner" to meet Onassis at avenue Foch. Shortly before the appointment, however, Onassis was summoned to Athens by Prime Minister Papadopoulos. Karr called Meyer on a bad line and told him that Onassis wanted Gratsos to take the meeting. Meyer did not question Karr's message, although he knew that it was Georgakis who should have been asked to deal with the banker in Onassis's absence.† Georgakis was chairman of the airline, and, as Georgakis later told me: "Protocol alone demanded it."[11]

---

* In 1964, Abu Jihad had accompanied Yasser Arafat to Peking, traveling on false passports under the respective pseudonyms of Galal Mohammed and Mohammed Rifaat.

† Was there a genuine misunderstanding because of the poor telephone line? Meyer later told Georgakis that if a Greek had given him the message, he might have asked him to repeat it; but he had no problems understanding another American, even on a bad line.

He believed that Karr had deliberately tried to exacerbate the tension that already existed between the Greeks at avenue Foch. Nobody could teach Karr—who had been a specialist in managing proxy fights in corporate takeovers*—a thing about the art of divide and conquer, the Olympic chairman ruefully reflected. "Karr sensed a vacuum in the organization and was determined to fill it," Georgakis said.[12]

Prematurely balding, with heavy-lidded dark eyes and a newly grown wide black moustache, Mahmoud Hamshari looked much older than his years. He had a habit of smiling briefly before he spoke; without the smile, his was an imperturbable face. He was a man who affected old-world manners; the fanaticism that defined him, the hatred that had molded him, were kept carefully masked in polite society. Nothing in his face would have given a hint of what must have been going through his mind as he appraised the value of Onassis's apartment and calculated the possibilities of the opportunity that lay before him.

According to Meyer, who was present throughout the first meeting at avenue Foch, the atmosphere was relaxed but businesslike. Gratsos, who knew something about the way Arabs liked to do business, waited for him to get to the purpose of his visit. The purpose, however, was not the one Hassner had earlier indicated to Meyer at the Plaza-Athenée.

He had, Hassner eventually began, speaking with the air of a man who deplored fanaticism, been approached by a *fedayeen*† terror group that was demanding a ransom of three hundred and fifty thousand U.S. dollars (about $3.5 million in present-day money) not to put a bomb

---

* In 1956, Karr made himself an instant authority on corporate takeovers with the publication of a book called *Fight for Control*. Much of the expertise that went into this book came from veteran proxy fighter Leopold Silberstein, president of Penn-Texas Corp., for whom Karr had worked and helped take over a company called Niles-Bernard-Pond. "It was a lesson the teacher would dearly regret giving, because in 1959 Karr joined with corporate raider Alfons Landa and engineered the takeover of Silberstein's company, which became Fairbanks Whitney, later renamed Colt Industries." (Roy Rowan, "The Death of Dave Karr and Other Mysteries," *Fortune*, December 3, 1979.)

† Freedom fighters: literally "men of sacrifice."

aboard an Olympic aircraft. He was acting as an honest broker, a facilitator, and did not know the identity of the terrorists who had contacted him through the Palestine National Fund, which looked after the finances of several terror groups, and whose officers had assured him of the determination and validity of the men he now represented.

Hamshari's pitch did not particularly surprise Gratsos. "We got blackmailed all the time," says Onassis's security adviser Miltiadis Yiannakopoulos. "We paid for the planes; we paid for the ships. It was impossible to operate without paying somebody money. Nobody called it blackmail. It was business. It was the way things worked in that world." [13]

When the Palestinian demanded money from Ari, therefore, it was nothing new. "But I was surprised because Ari was for the Palestinians," Yiannakopoulos told me. "He had given them money in the past. The deal was handled in Paris; I was in Athens, and was not involved. I was angry when I heard about it much later. When I mentioned it to Ari, he shrugged it off. He didn't want to talk about it. I always felt maybe he'd paid too much." *

Three-hundred-fifty-thousand U.S. dollars, Hamshari repeated, as if the modesty of the demand was in itself a sign of his own largesse. It was not an exorbitant sum when one considered what was at stake: the lives of the passengers, the reputation of the airline, the business that would be lost if one of its planes was blown out of the sky.

Hamshari was a pro, a man who had done his homework, and Gratsos respected him for that. He would inform Mr. Onassis of their conversation, he said, one businessman to another. Hamshari smiled his brief rictus smile. He would be in touch, he said.†

---

* Yiannokopoulos also told me of his own firsthand experience of Ari's business tactics. Interested in entering the shipping business himself, Yiannakopoulos once bought two tankers from Onassis for one million dollars. "Two days after [we shook on it], he calls me and tells me the deal will close at one million one hundred. I asked him, why the extra? He said, 'Oh well, come on, what is the difference between one million and one million one hundred?' So I made the deal for one million one hundred."

†At first, simple chronology appeared to contradict Johnny Meyer's claim that the first avenue Foch meeting between Hamshari and Gratsos had taken place in January or

*   *   *

In spite of all his kindnesses, the colonels were giving Onassis a difficult time in Athens.

Their internecine rivalries continued to bedevil the finalization of Omega, and he was tired and in a bad mood when he returned to Paris and heard the report of Gratsos's meeting with Hassner. Onassis did not regard the use of coercion as entirely inappropriate in business, and he had learned to live with the kind of gangster power behind the Palestinian's proposition, but the amount he was demanding was excessive. "We can't pay off these people with the tops'l sheet,"* he told Gratsos,

---

early February of 1968. "We know that Hamshari was in Paris soliciting funds for Fatah at that time, and one of several names he was using was Michel Hassner," says a former Mossad counterintelligence agent interviewed for this book. (Another name Hamshari used in France was Professor Maurice Hafez.) If Onassis gave Hamshari money before July 1968, it was because "Onassis *wanted* to contribute" to the cause, and not because his airline was threatened. There were three reasons for Mossad's thinking: (1) They knew that Onassis was "anti-Semitic and sympathetic to the Arabs." (2) Skyjacking was a PFLP (Popular Front for the Liberation of Palestine), not a Fatah speciality. (3) The first skyjack did not take place until *July 22* (when an El Al Boeing 707 flying out of Rome to Tel Aviv was hijacked to Algiers). Ergo, Hamshari could not have known in *January* what the PFLP planned to do more than six months in the future. Although Mossad do not dismiss Meyer's story, they question its timing. In order for Hamshari to have been able to use the PFLP's plans to blackmail Onassis, they believe that the meeting at avenue Foch described by Meyer must have taken place after, or much closer to, the first skyjack in July.

However, in an interview for this book, Bassam Abu-Sharif, who had been deeply involved in the PFLP terror campaign (a prime target on Mossad's hit list, he switched his allegiance from Habash's PFLP to become a leading moderate voice in Arafat's PLO after he was almost killed in 1972 by a Mossad parcel bomb), explained the apparent discrepancy in Meyer's story. Preparations for the first skyjack, he said, began in November 1967. "It was Wadi Haddad's idea. It was stunning. The way just a few men could take power against the many. It was breathtaking stuff. But we had penetrated Fatah's security, and we knew that they had infiltrated people into our organization. It would have been a miracle if they hadn't got wind of what we were planning," Abu-Sharif admitted. According to Patrick Seale (*Abu Nidal: The World's Most Notorious Terrorist*, p. 95. London: Arrow Books, 1993), the PFLP eventually became so penetrated by rival factions and foreign intelligence agencies that most of its plans had to be aborted.

* An old Greek shipping expression meaning to pay with a promise but no money (there is no tops'l sheet).

but neither would he pay the first figure that had apparently come into the Palestinian's head.

Hassner's true identity was soon discovered. Since he claimed to be an investment consultant with the Arab Bank, and Onassis was a friend of Abdul Hameed Shoman, a onetime haberdashery peddler who owned the Arab Bank, and whose son, Abul Majeed Shoman, ran the Palestinian National Fund, a couple of telephone calls told Onassis all he wanted to know about the Palestinian.

Onassis told Meyer, "Get the little fuck back here fast."[14]

Like Joe Kennedy, Onassis was at least a casual anti-Semite. "Beware of Jews," he had written to Ingeborg Dedichen in the 1930s, "and of being 'kiked.' "[15] He had never been persuaded by Winston Churchill's Zionism, and believed the Palestinians had been given a raw deal, first by the British, and now by the Americans. And almost on sight, he liked Mahmoud Hamshari. He liked his sly humor ("If I were not so deeply involved in politics, Mr. Onassis," Hamshari told him, "I would probably be a millionaire like you"[16]) and his lack of revolutionary rhetoric.

Both men had survived bloodshed in their youth. Exile and a sense of revenge had shaped their lives, and there is no bond more powerful than that founded on a shared sense of persecution. It was America that had forced their hands, compelled them to leave justice out of their reckoning. They agreed on many things, and it was not long before they started to like each other.

Neither Gratsos, Meyer, nor Karr attended that first meeting at avenue Foch or knew what they discussed, apart from blackmail, at that meeting, but the ground was laid for a deal that would change both their lives.

NINETEEN

A FAMILY WEAKNESS

**Politics is not the art of the possible.
It consists in choosing between the
disastrous and the unpalatable.**

—J. K. GALBRAITH, 1908–

Hamshari and Onassis got on well on a personal level, but the Palestinian was all business. The terrorist trade was the perfect place for him at that point. Arafat's people were still slowly recovering from the defeat of the Six Day War; Fatah was factional, undisciplined, confused, and ripe for someone like himself with an eye for the main chance. Nevertheless, Hamshari's demands were the least of Onassis's problems at that moment.

The Omega contracts remained unsigned, and rumors that Papadopoulos was going to switch the *junta's* support to Niarchos were rife in Athens. Nor was there much comfort in Onassis's private life; Tina's affair with Niarchos; Alexander's refusal to give up Fiona Thyssen, a woman with whom Onassis was himself more than a little taken; Jackie's stubborn reluctance to name the day, while continuing to be seen with a string of attractive men— Roswell Gilpatric, David Harlech, the photographer Peter Beard, and the architect Carl Warnecke among them—all added to the pressure on him.

"Even though he intended to marry Jackie, the fact that Tina was sleeping with Stavros bothered Ari tremendously," said Sir John Russell, who also believed that sexual ego was at the heart of his objection to Fiona Thyssen's affair with Alexander.[1]

Thyssen was aware of it, too. "It was not that I was sixteen years older than his son; it had nothing to do with the fact that I was not a Greek shipping heiress, the problem was that for the first time in Ari's life, Alexander had something that he wanted—and Ari knew he could never have," she told me. Just as she'd had too much dignity to be bothered by Tina's lies, so, too, did she have too much class to be seduced by Onassis's wealth. Yet her affair with Alexander saddened her as much as it angered Onassis. For although the relationship did not fit in with the image she had of herself, or what she wanted to do with her life, the harder she tried to leave, the more determined Alexander became not to let her go, and the longer the battle of wills went on, the more difficult it became for her to deny her love for him. "I liked being needed. I enjoyed teaching Alexander to think and to stand up to his father's tyranny," she said.[2]

A cool, sane presence in the midst of the emotional turmoil of the Onassis family, Thyssen watched with satisfaction as Alexander's perception of his father slowly began to change. "On that level I knew that I was a threat to Ari," she told me.[3]

Shortly after a particularly bitter row with Alexander, Onassis had lunch with Nigel Neilson at the Savoy Hotel in London. Neilson had been introduced to Onassis by Sir Winston Churchill, an old friend of his father's. A war hero who had been awarded the British Military Cross for gallantry, as well as the Légion d'Honneur and the Croix de Guerre (avec Palme) for his bravery fighting with the French SAS after the Allied landings, Neilson had been with Onassis since 1953, and was a confidant as well as his public relations man. Onassis told him that Alexander was pressing him to buy a helicopter to replace the Piaggio which was the workhorse of his Olympic Aviation taxi company in serving the Greek islands.

"Ari knew that the amphibian was coming to the end of its commercial life but he suspected that it was Fiona's idea to replace it with a helicopter. That was all he needed to be against it. By this time you couldn't mention Fiona's name without making him apoplectic," said Neilson, who regarded it as a family matter in which he did not want to become involved, and waited until Onassis's rage at Thyssen blew itself out.

"He wanted to have her legs broken, her looks ruined, her face pulped. He could express the most antisocial traits when he was in one of his snits," Neilson told me. But even the phlegmatic Neilson had been "momentarily unnerved" by the bitterness of Onassis's tirade against his son's mistress.[4]

Meanwhile, in Athens, not only were the colonels arguing amongst themselves about the spoils of their *putsch,* but Niarchos's belated arrival in the marketplace had brought an even brighter look of avarice and expectation to their keen soldierly eyes. Hurriedly playing catch-up, Niarchos had fastened onto Papadopoulos's deputy and ambitious rival, Colonel Makarezos, the *junta's* toughest critic of Omega.

Operating the way he did—buying off those who stood in his path, from bureaucrats to desert kings; dealing with those he could not buy in ways that would not bear scrutiny—nothing was ever straightforward for Onassis. Intrigue as well as corruption was the condition of every big deal he had ever done. And although he still marginally had the colonel's backing, he knew that Niarchos was quickly gaining ground. Some of his former brother-in-law's remarks made it clear that he knew many of his Omega secrets, too.

Onassis's fears were understandable. Almost nothing had gone right with Omega, and there was increasing uneasiness about the project within the development team. "Financially, Ari was about as far out on a limb as it's possible to get," Georgakis admitted.

Because of the close, almost incestuous relationships that characterized the jet set in the 1960s, Tina Blandford—a Greek educated in England, raised in America, and currently married to an English aristocrat—regarded Fiona Thyssen—a proud, independent Scot, and ex-wife of a German billionaire—with a profound sense of sibling rivalry.

It was not, therefore, maternal concern that compelled Tina to seek to break up Alexander's affair with Thyssen, but jealousy. That a woman only three years younger than she was—a woman who had once dated, vied for, and possibly slept with, the same men in their tiny milieu—was her son's mistress was an intolerable affront to her *amour propre.*

But what Tina did next would have consequences in ways she could never have imagined. As Georgakis told it, she called Onassis in Paris and told him that Niarchos was paying Thyssen "enormous sums of money" to sleep with Alexander in order to find out his Omega secrets.[5]

Onassis knew that Tina had a fertile imagination, and a Livanos aptitude for intrigue, but this time he believed her; probably because he knew that she was sleeping with Niarchos, and would be very well-informed about such matters—but also because it was what he wanted to believe.

For it finally explained why Thyssen appeared to prefer Alexander to him!

"Tina was a whore," snipes Onassis's former security chief, Miltiadis Yiannakopoulos, who was puzzled by Onassis's deference to his former wife.[6] Onassis could have had no illusions about her loyalty, and might even have encouraged her infidelities, because they had excited him during their marriage. (The English writer Alan Brien recalled an episode that occurred when he was a guest at the Onassis villa in the South of France in the 1950s. Unable to sleep, he went in search of a book to read, and he heard Onassis screaming at Tina, *"You whore! You whore!"* The door to their suite was ajar; Brien saw shadows of "a man hitting out, the shape of a woman covering up." Just as he'd decided to intervene, "they began kissing passionately, and withdrew to the bedroom, obviously to fuck."[7])

But if Onassis had given Tina a taste for betrayal, as well as a little masochism in her sexual relations, it seems to have paid off. Shortly after her claim that Thyssen was working for her lover, Niarchos, there was not a Niarchos office, house, love nest, yacht, or car Onassis had not had bugged, or a phone that was not tapped.[8] "If we heard that Niarchos was looking for a new maid, she would be ours before she arrived," Yian-nakopoulos told one reporter. "It would not cost much, but it would be a sum she would never have dreamed about."[9]

Onassis summoned his son to Paris and confronted him with his mother's accusations, and ordered him to dump Fiona—"or I'll get rid of her for you," he yelled.[10]

His father's parting shot troubled Alexander. "The old man never makes empty threats," he told Thyssen, and urged her to take special care of herself when he was away. (Although they spent every weekend together, they never lived together; he worked in Monte Carlo, and she kept her home in Morges, near Lausanne.)

"I knew that Ari was an extremely dangerous person; he would stop at nothing to get his way. There was something just awful in the air. One hesitates to call it evil, but that is what it was, something quite evil," Thyssen would later tell me.[11]

In Paris, Onassis continued to deal personally with Mahmoud Hamshari. They met at avenue Foch, and sometimes at a café on the Place de la Sorbonne, a small square lined with cafés and bookshops next to a side entrance of the Sorbonne. When the meetings were at the café, Meyer waited at a bar across the square until Onassis signaled for his car.

One reason the talks dragged on was that Hamshari would sometimes simply disappear. The fact that he was spending time in Los Angeles became apparent only when he complained that the LA smog seriously affected his sinuses and gave him terrible headaches. Karr referred him to a physician he knew in LA named William Joseph Bryan, Jr.

Bryan, whom Karr had met during his time as a producer at Metro-Goldwyn-Mayer, had a medical and hypnotherapy practice called the American Institute of Hypnosis on Sunset Boulevard. He had success-fully treated many Hollywood stars for their drug and alcohol abuse through hypnosis. Karr had taken Aldo Ray to him to get the actor off the booze, during the filming of *Welcome to Hard Times*. Bryan had also been the technical adviser on the filming of Richard Condon's classic Cold War thriller, *The Manchurian Candidate*.

Karr gave Hamshari Bryan's number and thought no more about it.

*   *   *

March 16 was a Saturday, and Onassis was schmoozing with Karr and Meyer in the bar of the Georges V in Paris when they heard the news that Bobby Kennedy had declared his candidacy for the presidency. The idea of another Kennedy in the White House worried Onassis far more than a possible Nixon victory. "If Bobby wins," he said, "he'll want to destroy the colonels just to fuck me." [12]

He had been drinking since lunch time and could not have been altogether sober when, later that evening, he called Jackie in Mexico and told her that he wanted to bring forward the date of their wedding.*

Although Jackie was probably no fonder of Onassis than she was of several other lovers at that time, the pure ferocity of his greed excited her in a way that love alone probably never could. If she felt any love, excepting the love she felt for her children, it was for money. She craved the lifestyle Onassis held out and the security he promised. He was an older man, and she loved older men. But he possessed one of the most remarkable tax-free cash flows in the annals of plutocracy, and she loved that most of all.

"Actually, she was far keener on the idea (of marriage) than Ari was," says Lilly Lawrence, the daughter of Dr. Reza Fallah, head of the Iranian Oil Syndicate, and one of the few people in whom Onassis had confided that he planned to marry Jackie. "Mrs. Kennedy would have done anything not to lose him." [13]

Nevertheless, Onassis was tainted irredeemably in American eyes, and the marriage was going to upset a lot of people. But in November, John Kennedy would have been dead five years; in November, too, there was a real prospect of there being another Kennedy in the White House to soften the lingering sense of national loss, and there probably wouldn't be a better time for her to cast off her Holy Widow mantle than now.

---

* Onassis told the author that he had called Jackie in Mexico shortly after he heard that Bobby had declared his candidacy. However, Jack Newfield (*Robert Kennedy: A Memoir*, p 229. New York: E.P. Dutton, 1969) claims that when Bobby marched in the St. Patrick's Day Parade in New York that same afternoon, Jackie had blown a kiss to him from an open window.

But first she had to break the news to Bobby himself, and her apprehension was not misplaced. Still her occasional lover (from whom she continued to collect a useful $50,000 a year), brother-in-law, and—since Joseph Kennedy's stroke four years earlier—head of the Kennedy clan, he was not a man she wanted to cross.

Nevertheless, she was now thirty-eight years old. In no time she would be on the edge of middle age. "Psychologically, it will be hard for Jackie to see Ethel in the White House as another Mrs. Kennedy," said the French writer and Kennedy family friend Paul Mathias. "What will she be then, the Queen Mother?"[14] The question went to the heart of her dilemma. For nearly five years her widowhood had been the central fact of her life. The embodiment of a nation's grief, she had played her role of Holy Widow faultlessly (aside from the Manchester authorized biography incident). She'd had lovers, more than people ever suspected, but, with the possible exception of Bobby, she had never fallen in love.

"When death ends one dear relationship, it often creates another sweeter still," she had written to Onassis in an earlier attempt to explain why she would always be devoted to Bobby. He would always be more to her than her husband's brother; there was a time, she wrote with touching but rather economical candor, when "Bobby was more to me than life itself." *

The truth was, Bobby still had an extraordinary power over her. Only he could have persuaded her "to put on my widow's weeds and go down to LBJ's office and ask for tremendous things," she had told one friend[15]; she would put her "hand in the fire for him," she said.[16] She knew that he would not allow her charismatic widowhood—the biggest political asset he had—to slip through his fingers at the moment he needed it most.

---

* Nobody will ever know how many letters Jackie wrote to Onassis between 1963 and his death in 1975. According to his sister Artemis, who read to the author the excerpt quoted here, there were "so many . . . hundreds." Some, she said, contained no more than a few lines of affection, private thoughts, or simply details about some domestic matter. Christina Onassis is believed to have destroyed all of them shortly after her father's death in 1975.

But while Jackie fretted and stalled, Onassis decided to take the ini-
tiative. He summoned David Karr.

Karr wrote President Lyndon Johnson on March 26 and informed him
that he would be making one of his "rare trips back home" and staying
with his old friends, Drew and Luvie Pearson. "If you can spare a few
moments, I would like to drop by and pay my respects. As you may
know, I've done rather well here in the investment banking business. Perhaps I can add a positive note to your vast fund of knowledge on our
recurring problems in the international financial area." [17]

Johnson agreed to meet Karr and Drew Pearson on April 3.

Before Karr left Paris, Onassis gave a small cocktail party for him
at the Trémoille hotel. Among the guests were the lugubrious Paul
Bougenaux (who smiled only at funerals, according to one of his man-
agers), Johnny Meyer, myself, and a sprinkling of Madame Claude
girls.

It was at this gathering that Onassis made his now-notorious remark
to me about Jackie needing "a small scandal to bring her alive. A pecca-
dillo, an indiscretion."

The world, he said, loved fallen grandeur. [18]

According to the White House log, David Karr and Drew Pearson
entered the Oval Office at 1:56 P.M., on Wednesday, April 3, and stayed
for fifty minutes. No notes were taken at this off-record meeting. But
whatever Karr told Lyndon Johnson that day was sufficiently compelling
to have detained him for nearly an hour on one of the longest and most
critical days of his presidency.*

---

* Earlier that morning, eleven months after LBJ had escalated the Vietnam War, Hanoi
radio had announced that "Hanoi is ready to talk." The report had precipitated a
stream of meetings, and conferences with Lyndon Johnson's most senior aides and advi-
sors, which had gone on nonstop for the rest of the day and into the night. According
to the White House diary, the President's first appointment was at 8:00 A.M.; his last
meeting did not end until 12:45 the following morning. At 11:09 that evening his
daughter Lucy called "asking her father if he wanted dinner sent over." He declined.

At the height of the Cold War, revelations about Omega and Onassis's plans to buy crude from the Soviets, as well as the fact that he was negotiating with Palestinian terrorists to save his airline, would undoubtedly have held the president's attention. But Karr would have left his choicest piece of gossip, and the primary purpose of his visit—to "let the cat out of the bag among the pigeons," in Costa Gratsos's symphystic phrase—until the last:

Aristotle Onassis and Jackie Kennedy were going to be married.

This would have been raw meat to Johnson. Three days before, he had announced that he would not run for another term, and he was still seething at the Kennedy camp's slur that Bobby's challenge had driven him back to the ranch.

When Johnson planned to play this card we shall probably never know. But shortly after Karr's visit to the Oval Office, Eugene McCarthy saw Johnson, and when he brought up the subject of Bobby's presidential run, *"The president said nothing: instead he drew a finger across his throat, silently, in a slitting motion."* [19]

It was unreasonable to expect Jackie to go on pretending to be a devoted widow for the rest of her life. But her sense of class, which had made John Kennedy look so good, wouldn't be worth a dime to Bobby if she could no longer be presented as his brother's widow, but only as Mrs. Aristotle Socrates Onassis, the wife of the crook Bobby himself had banned from America when he was attorney general.

Nevertheless, Jackie and Bobby had had time to reflect and were able to talk calmly when they eventually met at her apartment in New York to discuss the situation. Although he was appalled at her judgment, he could not afford to lose his temper with the woman who had created the Camelot legend, the myth on which so much of his hopes now rested. So instead of being angry, he was affectionate and conciliatory; instead of being coquettish, she was shrewd and practical. Marriage wasn't necessarily imminent, she told him, but it was a strong possibility in her mind; she did not love Onassis, love didn't come into it, but each had something the other needed; and after five years of widowhood, she had to decide what to do with the rest of her life.

Still unaware of the time bomb that was ticking away in that part of Lyndon Johnson's mind in which he kept old scores that had to be settled, Bobby asked her to understand the embarrassment it would cause him if news of their affair leaked out in the middle of his campaign. "For God's sake, Jackie, this could cost me five states," he told her, and alluding to Onassis's earlier affair with Lee, he added: "I guess it's a family weakness." [20]

But what did he want her to do?

If she agreed to do nothing that would put his political message into eclipse, he promised he would try to adjust to the idea of "the Greek" after the election. An expert in Kennedy lore—whatever the moment's purpose, everything must serve it—Jackie knew that this was the best offer she was going to get, and agreed to suspend her retirement from public life to campaign for him. She also promised to wait until after November before making any public announcement about her future.

That evening she relayed Bobby's offer to Onassis in Paris. As Georgakis remembered it, "Ari knew that Kennedy was stringing them along. He knew that in November Bobby would hold all the cards. 'He'll shuffle me out of the deck once he's in the White House,' he said." [21]

Johnny Meyer disagreed. Jackie would have outlived her usefulness to Bobby by November, he believed. "I said, 'Bobby won't give a shit what she does after the votes are counted,' " he later told Brian Wells. [22]

Onassis had always taken what he wanted from life, and for the first time, he had come up against a younger man who was as ruthless as he was. Bobby saw him as "the rich prick moving in on his brother's widow woman," as he crudely but precisely defined the problem. [23]

Meyer stuck to his guns. In November—whether Bobby won or lost, whether Bobby liked it or not—Jackie would be a free woman and able to make her own choices.

And in November, Onassis said grimly, Bobby would still be "the same sonofabitch she said she'd put her hand in the fire for." [24]

Meanwhile, Jackie told Onassis she feared that Bobby's luck was running out. Onassis thought she meant that he might not win the Democratic nomination for the presidency. "He said, 'She's talking through her hat,' " Georgakis recalled of Onassis's attitude. "He believed that only an act of God could stop Bobby."[25]

Nevertheless, as Bobby's campaign rolled on, Jackie began to express her doubts more openly and more passionately. At a dinner party in New York, she told Arthur Schlesinger: "Do you know what I think will happen to Bobby? The same thing that happened to Jack . . . There is so much hatred in this country, and more people hate Bobby than hated Jack . . . I've told Bobby this, but he isn't fatalistic, like me."[26]

TWENTY

MISSING PIECES

Who is the guilty party?
The one who commissions the job,
or the one who takes it on?

—FRIEDRICH DURRENMATT (THE EXECUTION
OF JUSTICE), 1921–1990

In late March 1968, a morning of pale spring sunshine, Johnny Meyer accompanied Onassis to the café in the Place de la Sorbonne for another meeting with Mahmoud Hamshari, then walked across the square to sit at a table outside a bar and wait. His exclusion from the talks deeply hurt the man who liked to call himself Onassis's Falstaff. But Hamshari had cultivated a detachment which disconcerted Meyer, who suspected that Hamshari did not trust him because he believed he was a Jew, which Meyer denied.*

---

* He was punctilious in pointing out that his name was Meyer, not Meyers, the more common Jewish name. According to a certified copy of Record of Birth in the Office of the City Clerk of Fall River, Massachusetts, however, he was born on July 18, 1906, son of William Nathan Meyers, an advertiser from Cleveland, Ohio, and Ella Meyers (nee Ella L. Holmes).

Settling down for another long wait, Meyer opened a copy of the *New York Herald-Tribune,* and began marking stories that interested him or that concerned people he knew. Later he would clip the items he thought would interest Onassis and give them to him, with his comments scrawled in the margins. Sitting at a sidewalk café in the spring sunshine in Paris was not a bad way to earn a living, and he had blessed providence every morning of his life for guiding him into such rewarding work. But this morning he felt restless. From the moment he met Hamshari in the bar of the Plaza-Athenée in January, the idea of throwing in his hand had been on his mind. He was a rich man, not by Onassis's standards perhaps, but he had once struck oil in Wyoming and turned a $35,000 investment into a million. He had property in New York, a home and an attractive wife in Florida, as well as an apartment and a mistress in Paris, and on days like this the temptation to call it a day was almost irresistible. He had met a lot of unsavory people while working for Howard Hughes, but he'd never met anyone he distrusted more than Hamshari, he later told his old Hollywood friend and Palm Beach neighbor, Ernie Anderson.

"Johnny was major league pissed at the way Dave Karr tricked him into introducing the Arab guy [Hamshari] to Ari, and was then shut out. It hurt his pride," said Anderson, who, like Meyer, enjoyed a legendary reputation as a cynical PR.[1]

But, apart from the conspiratorial natures of Onassis and Hamshari—whose life was as full of missing pieces as Ari's—there was a very good reason why Meyer had to be cut out of the loop. For as well as being temperamentally attracted to plots, Palestinians and Greeks are often great games players; and when the two men met at the café in the Place de la Sorbonne that morning—ostensibly to discuss the blackmail of one by the other—a threshold had been crossed. They had embarked on what must have been the game of their lives.

They continued their discussions behind closed doors at avenue Foch; they talked through the afternoon and evening, and into the early hours of the following morning. Meyer had always admired the way Onassis could string out and confuse a deal to his advantage. "Ari could fuck up a two-car funeral if it was to his advantage to do so," he said.[2] But blackmail, in his experience, was not a subject of exhaustless debate, and he couldn't figure why the talks had dragged on so long.

Here we must backtrack a couple of weeks to an Israeli raid on a guerrilla camp at Karameh in Jordan where the outnumbered, poorly equipped Palestinians had forced an Israeli retreat. Seen as a heroic victory, money poured in from Palestinian sympathizers around the world at such a rate that a battalion of tellers had to be recruited to count it. The war had suddenly become big business; Arafat's picture appeared on the cover of *Time;* five thousand volunteers swelled the Fatah ranks.

Commuting under various aliases* between Paris and Los Angeles, Hamshari appears to have become a sidelined figure in the organization, his assignment a forgotten sideshow to the real business in Damascus. Nevertheless, he was enjoying the autonomy he had been given by Fatah's intelligence chief Abu Iyad—shortly after Hamshari's proposal the previous June that Fatah "kill a high-profile American on American soil"— to run Fatah's first covert western operation, and having adjusted to the good life in the West, he was apparently in no hurry to return to Damascus.

Although we shall probably never know until the archives of Fatah are opened to public scrutiny, and perhaps not even then, whether what happened next was a sanctioned operation or whether Hamshari was off on his own (terrorism will always be a grab-bag of opportunities for those smart or daring enough to take them), the circumstantial detail and known facts prove beyond almost all possible doubt that he had entered into partnership with Onassis to carry out his proposal—made and apparently disregarded at the Fatah meeting in Damascus in the shocked aftermath of the Six Day War—to assassinate a high-profile American on American soil.

---

* According to a reliable PLO intelligence source, Hamshari used at least four aliases and several occupational covers in Los Angeles and Paris before he assumed his own identity when he later became the official Palestinian representative in France.

Now Hamshari had put a name to his target:

Robert F. Kennedy.

Some might dismiss the Fatah leaders' apparent surprise and anger, and the reprisal they would inflict on Hamshari when they later learned of the enormity of his coup, and how much Onassis had paid for his services, as the cynical response of collaborators who wanted to remove a witness to their own complicity. But there is no evidence for thinking so, and, as we shall discover, a good deal to the contrary.

The Paris student revolt of May 1968 was approaching, and there was a sense of unrest in the air. "That's why I love Paris so much," Onassis told Meyer, when they passed armed riot police waiting in vans parked up in a side street, prepared to break up a student march against the war in Vietnam. "The French understand that a little violence applied at the right time can solve a lot of one's problems."[3]

Astonishingly, nobody had yet told Olympic Airways' chief executive Yannis Georgakis about Hamshari's threat to the airline. In fact, the first time Georgakis heard the Palestinian's name was from a friend at the Israeli embassy, who also informed him that Hamshari had been under surveillance by Mossad since his arrival in Paris, and that he had been meeting Onassis.* "I said, 'Ari's in the oil business; he meets lots of Arabs,'" Georgakis would later recall of the conversation. "My Israeli friend said, 'This Arab is not an oil man. He is a very senior Palestinian terrorist. A most dangerous fellow.'"[4]

Although he was angry at being kept in the dark, Georgakis said that it was not at all remarkable that Onassis was seeing a Palestinian—even one who was a terrorist. His sympathies for the Palestinians were understandable: They had been expelled from their homeland, just as he and his family had been expelled from Smyrna in 1922. Georgakis told the embassy man that Hamshari had probably been soliciting funds, and the subject was dropped.

---

* Although Georgakis declined to name his Israeli contact, it was probably ambassador Walter Eyton, whom he had known for some years.

Nevertheless, Georgakis's explanation was like a defense lawyer's automatic rebuttal that the prosecution has no case. In fact, he was deeply troubled by the Israeli's story, and although he still had no reason to connect the meetings between Hamshari and Onassis with the airline, he felt, like Meyer, a sense of unease about his exclusion.

Onassis had an almost congenital inability to do anything simply if there was a devious way of doing it. Georgakis claimed that by the spring of 1968, 88 avenue Foch was beginning to resemble a "Balkan court of the previous century . . . "spying, intrigue, and accusation" abounded, he told American reporter L.J. Davis.[5] This was an exaggeration, probably colored by his subsequent discoveries of what was being plotted between Onassis and Hamshari at that time.

It was in this atmosphere that he made the extraordinary decision not to warn Onassis of the Israeli's interest in Hamshari, and the fact that he was a terrorist.

A week or so after Georgakis's meeting with his Israeli contact in Paris, Onassis took him to lunch at the Hotel Grande Bretagne in Athens, where Onassis finally told him about the Palestinian's threat to blow up an Olympic plane. But Georgakis need not worry because he had dealt with the problem himself, he assured him. However, he needed $200,000 in cash from the Olympic coffers to pay "the first installment" of the protection money.[6] Future payments would be arranged "off the books" and channeled through his Panama corporations, he assured the startled Georgakis.

Such casualness might have been acceptable in the 1950s when Onassis ran the airline "between drinks at Maxim's,"[7] and his brother-in-law—the good doctor Theodore Garofalides—was its chairman, and as free with the petty cash as he was with his prescriptions. But Georgakis liked to run things according to rule; he had been hired to sort out the management shambles and put the airline on a proper business basis, and he was not, he told Onassis, prepared to preside over a company that was conducted on such "a *per incuriam* basis."[8]

According to Onassis, Hamshari wanted $1.2 million to guarantee the airline's safety.* Although Georgakis did not know it at the time, this was more than three times the figure Hamshari had asked for at the meeting with Gratsos at avenue Foch two months earlier. "I said, 'How much are the other airlines paying, Ari? Don't you think we should find out?' " Georgakis later recalled the conversation.[9]

None of the other airlines would admit that they were being black-mailed, Onassis told him—and neither must they. "To admit that we were paying protection money would raise questions about Olympic's safety, and that would be bad for business, and of course he was right," Georgakis agreed. And although he thought that Onassis had behaved badly in not confiding in him until now, he agreed to provide the initial $200,000, on condition that in the future he would be allowed to deal with Hamshari himself.[10]

"Ari said, 'What's the point? The deal's done. There won't be a next time,' " recalled Georgakis, who wanted to reply he no longer trusted Onassis's judgment; he now "made a mess of everything" he touched.[11]

Onassis flew to New York, where he handed $200,000 in cash—packed in brown envelopes and carried in a Saks shopping bag to his chauffeur, Roosevelt S. Zanders, and told him to take the money to an apartment in the United Nations Plaza on First and Forty-Ninth Street, an address popular with UN diplomats and politicians.†

"Mr. Onassis and I went back a long way together. I took the first Mrs. Onassis to hospital when she was expecting Christina," Zanders told me, recalling the trust Onassis placed in him with pride. "He gave me a Saks bag. He said, 'Mr. Zanders, there's two hundred thousand dollars in this bag.‡ Don't leave it with the desk captain. Take it straight up

---

* According to Johnny Meyer, the figure was $3.2 million. However, allowing for Meyer's hyperbole and world inflation, Georgakis's figure is the one I will stick with. Significantly, however, in both accounts the $.2 million did not change.

† Bobby Kennedy and Truman Capote also kept apartments there.

‡ Unaware of its dark purpose, Zanders' never forgot the small fortune Onassis handed him that day. "Aristotle Onassis trusted me with two hundred grand. Maybe I should

to the apartment yourself.' He said I wouldn't get a receipt. 'But I know I can rely on you,' he said.

"I said, 'Two hundred grand's a lot of money, Mr. O. I hope I don't lose it.' He said, 'Lose it, I'll kill you!' " [12]

That evening, Jackie's friend (and soon to be Ari's lover), Joan Thring, gave a small dinner party for Onassis and Jackie in New York. It had been Jackie's idea to have the dinner at her own apartment. One of the few people outside the two families who knew that Jackie had finally told Bobby and the family that she planned to marry Ari, and aware of the need for discretion, Thring had asked Jackie to draw up the guest list herself. Jackie had wanted only three other people at the dinner: her mother, Janet Auchincloss, and Bobby and Ethel Kennedy. There was, understandably, no invite for Lee. But if it was her hope that Bobby and Ethel—now that she had agreed to Bobby's terms for the marriage—would find a place in their hearts for Onassis, it was not, Joan thought, a rational expectation. "It was obvious that nobody was pleased about what was going on," she told me.[13]

Janet Auchincloss had been no mean social climber herself in her day, and she must have understood Onassis's determination to elevate himself above the ordinary marketplace and make a niche for himself in history by marrying the president's widow. And however much she disapproved of Onassis as a human being, however much his charm grated on her nerves, she otherwise might have been proud that her eldest daughter had, for the second time in her life, so punctiliously followed her advice to marry money. Nevertheless, as a mother, she could never forgive Onassis for dumping her baby, Lee, when her more collectable sister became available, and she could barely utter his name without grimacing.*

---

have that on my gravestone," he told the author. In his obituary in the *New York Times* (May 26, 1995), Robert McG. Thomas, Jr., reported, "He made a $200,000 cash delivery for Aristotle Onassis."

* Staying at Claridge's some years earlier, Janet had been told that Lee was visiting Onassis at the hotel. Wishing to talk to her daughter, she had gone to his suite, to be

Ethel spent the evening in a rigor of glacial forbearance, and the few exchanges between Bobby and Onassis were also no more than icily polite. Bobby knew what "the Greek" had on him, of course. He would not have forgotten Onassis's answer when he had tried to stop him from seeing Lee Radziwill five years earlier: "Bobby, you and Jack fuck your movie queen and I'll fuck my princess."

There was so much history between them, so much hatred and revenge still to be wreaked, that neither of them dared relax.

Thring felt sorry for Jackie; she felt sorry for anyone unlucky enough to get involved with the Kennedys, she said: "They were all quite frightening, quite scary, all of them, absolutely ruthless," she said, appalled by "the way they were fighting to keep control of Jackie." [14]

But their cause was already lost.

The $200,000 in bills that Roosevelt S. Zanders had delivered that afternoon to an apartment in the United Nations Plaza, for which no receipt was asked or given, had already settled the matter.

---

greeted by Onassis in his dressing gown. "She was shocked by his dishabille at an hour when a proper gentleman would be savouring his pre-luncheon martini," a friend would later tell the story. She demanded to see her daughter. "And who exactly is your daughter, may I ask?" Informed that she was Princess Radziwill, he said: "In that case, madame, you've just missed her."

TWENTY-ONE

AN INEXHAUSTIBLE SUPPLY OF EXCESS

Somebody is going to try to kill you.

—ROMAIN GARY TO ROBERT KENNEDY

The following morning, Onassis sent Joan Thring roses and an invitation to join him on his yacht in the Caribbean when he returned from Las Vegas, where he had a meeting with the corporate raider Kirk Kerkorian. According to David Karr, Kerkorian was secretly eyeing Metro-Goldwyn-Mayer, and might need to unload his Western Airlines stock in a hurry to finance a raid on the ailing movie company. It would be an opportunity for Onassis—who had for some time been trying to acquire a U.S. domestic carrier to extend Olympic's routes across America—to get a very mean deal.*

---

* It was a perfect example of how Karr operated, said Yannis Georgakis: "David knew that Kerkorian had borrowed heavily against his Western Airlines stock, and would be stretched if he didn't get control of MGM by the time the stock started to rise. Ari wanted a U.S. carrier, putting him together with Kerkorian was very clever."

Kerkorian acquired MGM—with, many believed, Onassis's backing;* Onassis, however, never got Western Airlines.† But did Onassis have another reason for going to Las Vegas, a town he loathed—indeed, Rainier's desire to turn Monte Carlo into a Vegas-style operation had been the starting point of their disastrous rift—in May 1968?

Why did he, for example, also visit William Joseph Bryan, Jr., the hypnotherapist to whom Karr had sent Mahmoud Hamshari to treat his migraine headaches in Los Angeles?

Rupert Allan learned of their appointment at the Riviera hotel when Meyer, unable to reach Karr in Paris, called him and asked if he had Bryan's unlisted telephone number. "He thought Bryan had treated Marilyn (Monroe), and I might have his number," Allan recalled.[1] But Monroe had used so many physicians, Allan didn't know whether Bryan had treated her or not. He got the number the next morning from a contact at Fox and called Meyer. "I said, 'Johnny, you wake me up in the middle of the night, Ari better be seriously green around the gills.' He said: 'He's got serious insomnia, Rupert. It keeps *me* awake nights. Now it keeps *you* awake nights. It's contagious as hell.' "[2]

Another of Bryan's specialities, however, was the treatment of sexual dysfunction in the aging male; his ability to perform the ultimate apotheosis attracted clients from all over the world. And, in spite of Onassis's boastfulness about his sexual prowess, as he contemplated marriage to a woman nearly thirty years his junior, a woman in her sexual prime, and, if his own testimony is to be believed, voracious in bed, he might have felt he needed all the help he could get.

---

* *Variety* editor Peter Bart, who was at that time a senior vice president of the studio, wrote that Wall Street was awash with rumor about the source of Kerkorian's funding: "It was Onassis money, it was Arab money . . ." (Peter Bart, *Fade Out: The Calamitous Final Days of MGM,* p. 31. New York: Simon & Schuster, 1990.)

† "What happened in Las Vegas was never entirely clear to me. Ari told me we'd got Western. Wonderful news. A few days later, it isn't quite so wonderful. Then it isn't wonderful at all: The deal is off. Nothing was ever straightforward when David Karr was involved. He always had another agenda," said Yannis Georgakis.

Nevertheless, this episode would probably have been forgotten had Allan not bumped into David Karr at the Pierre Hotel in New York in the spring of 1977, as we shall eventually see.

Dressed in a crumpled white suit that Meyer thought made him look like an ice-cream salesman, Onassis left Las Vegas and returned to New York with Johnny Meyer and actor Cary Grant on a jet belonging to the Fabergé cosmetic company. After picking up Joan Thring, they continued on to Miami, where they joined the *Christina* and headed for the Caribbean.

Cruising with celebrities brought out the best in Onassis. His guests fell under the spell of his charm, his stories, his bawdy jokes, and his outpouring of Onassisian wisdom: "I approach every woman as a potential mistress . . . Beautiful women cannot bear moderation: they need an inexhaustible supply of excess . . . I get up every day of my life to win." Each evening a piano–violin duo serenaded the guests while white-gloved waiters served dinner, after which they sipped champagne and danced on deck by the light of Chinese lanterns. Each morning another island would be waiting for them to visit. "It was magical," remembers Thring. "It was as if the real world didn't exist." [3]

In the real world, however, the dominoes were falling Bobby's way. Following his primary victory in Nebraska, the polls predicted further success for him in Oregon and California.

Nevertheless, a sense of foreboding was deepening around him. The French novelist Romain Gary, who would commit suicide three years later, and like many suicides had an instinct for impending death, told Bobby when they met shortly after Martin Luther King's murder in Memphis, "Somebody is going to try to kill you." (He was far more blunt when he talked to Bobby's campaign aide, Pierre Salinger: "Your candidate's going to get killed." [4]

There were no guarantees against assassination, Bobby replied; either luck is with you or it isn't. "I am pretty sure there'll be an attempt on my life sooner or later. *Not so much for political reasons . . .*" [5] Some day, he believed, people would no longer be able to mention "the Kennedy assassination" without specifying which one,[6] and reporters covering his campaign shared the premonition: He had the stuff to go all the way, but he wasn't going to make it. "The reason is that somebody is going to shoot him," *Newsweek*'s John Lindsay told Jimmy Breslin. "I know it and you know it . . . He's out there now waiting for him."[7]

Nobody felt this more deeply than Jackie. Obsessed with the amount of violence in the world, she had grown pessimistic about the fact that nobody even seemed to care. "Of course people feel guilty for a moment. But they hate feeling guilty. They can't stand it for very long. Then they turn," she had told Bobby after she returned from the funeral of Martin Luther King, Jr, in April.[8]

At the Caribbean island of St. John, Cary Grant, Johnny Meyer, and the rest of the guests, with the exception of Joan Thring, left the *Christina* to fly back to New York. Jackie arrived the following morning. She was piped aboard as if she were visiting royalty. "I had no idea that she was coming. She hadn't said a word to me about joining us, and neither had Ari," Thring complained later. "I was upset because I suspected that I'd been taken along to act as a beard. There were always little boats floating around the *Christina,* and Ari was terrified that somebody would get a picture of him and Jackie alone. He kept telling me, 'For Chrissake stick close; don't leave her side.' I thought that he was probably more afraid of Maria finding out than what the newspapers would make of it."[9]

But it was already too late to stop their tryst from reaching the ears of another deeply interested party. And on May 17, a few hours after Jackie stepped aboard the *Christina,* President Johnson finally acted on Karr's tipoff that the former First Lady planned to be the second Mrs. Onassis, with a memorandum to J. Edgar Hoover asking for a full FBI report on her groom-to-be.[10]

Jackie remained aboard the *Christina* for six days and nights. Each afternoon she slipped into Onassis's stateroom to make love and pursue the awkward questions their marriage would raise. Onassis shaved twice, even three times a day (probably because his beard was grey, which made it more difficult for him to carry off his claim that he was only sixty-two; he was, in fact, two years short of his seventieth birthday). In the evenings, Jackie encouraged her lover's reminiscences, allowed him to light her cigarettes, confessed to him her likes and dislikes. "Do you mind taking off your glasses?", she asked one evening. "Dark glasses are very forbidding on a person at night."

Jackie and Joan Thring got on very well together. "I was slim, and Jackie was always very slim; we both had our hair pulled back, and tied behind," Thring recalled. "Ari loved that. He called us his two lovely boys. He started off liking boys, of course. We must have seemed an odd *menage à trois*." [11]

But although the cruise appeared to have all the elements of an idyll, Jackie and Onassis were sailing into seriously choppy waters. For whatever Bobby had promised Jackie to her face, according to Thring—who had hosted the strained *en famille* dinner party in New York and knew the true score—he was assuring others that she would "marry Onassis over his dead body." [12]

Nevertheless, Thring—an attractive, straight-talking Australian who was Rudolf Nureyev's personal assistant at that time—knew that "there wasn't a chance in hell Ari would let a little thing like Bobby stand in his way. He and Jackie would vanish in the afternoons—presumably to discuss the small print. It was, after all, a business arrangement. Ari would get what he wanted, which was to get back into America in a big way, and Jackie would get heaps of money." [13]

On the evening Jackie flew back to New York, and the *Christina* began the slow voyage back to Europe, Onassis and Thring became lovers. Each afternoon they made love in a different stateroom. Some thirty years younger than Onassis, Thring wondered where he got his energy. After their lovemaking, he went back to work; keeping in touch on the radio-telephone with his people in Europe, New York, and Buenos Aires; talking to the captains of his tankers at sea, and to his bankers in Zurich, Panama, and the Dutch Antilles, wheeling and dealing, moving and hiding his money. Yet at dinner he was always fresh, prepared for the evening ahead. Dinner invariably began with Cristal champagne and Beluga caviar—a mixture which he said made him extremely horny. Some evenings he'd get drunk; other evenings no matter how much he consumed it didn't seem to affect him at all.

After dinner, they sat at the stern in deep sofas, listening to Maria Callas recordings. It was extraordinarily poignant to hear the beautiful voice of the woman who had been Onassis's devoted mistress for so many years, carrying across the empty ocean in the vast darkness. But after each aria, Onassis removed the disc from the player, solemnly broke it in two, and hurled the pieces into the ocean.

It occurred to Thring that it was his way of breaking with the past.

Or maybe he was going mad.

But as the *Christina* moved slowly back to Europe, and Onassis's declared intention to make love to her in every stateroom before the journey was over proceeded to plan, it seemed more and more to Thring as if Onassis was waiting for something to happen—and was determined to be someplace else when it occurred.

The California primary had been a great victory for Bobby.

Around midnight in the Embassy Room of the Ambassador Hotel in Los Angeles on June 5, 1968, he thanked his supporters with a mixture of humor and gravity. "We are a great country, an unselfish country, and a compassionate country . . . I think we can end the divisions within the United States, the violence," he told them with a poignancy that was still unimaginable.[14]

Surrounded by a clutch of aides, hotel employees, and newsmen, with his wife Ethel a few yards behind, and with the cheers still ringing in his ears, he left for a press conference in the Colonial Room on the other side of the hotel. The route they took, from the stage to an anteroom and into the service corridors, led them straight to a narrow serving kitchen that was to be his place of execution.

As the senator approached, a dark, slim young man stepped from behind a tray rack. "Kennedy, you son of a bitch," he shouted as he raised a .22 revolver to the senator's head and squeezed the trigger.[15] The fatal first hollow-point bullet exploded through the right mastoid bone, disintegrating into the right hemisphere of Kennedy's brain. Two more shots struck Kennedy's right armpit as he fell to the floor.[16] The assassin continued firing, wounding five other people as Kennedy aides wrestled him down onto a steam table where he was held until police arrived.

"I did it for my country," the attacker cried out.*

Onassis heard the news at about ten o'clock on the morning of Wednesday, June 5, 1968, while having breakfast in the middle of the Atlantic Ocean. Bobby wasn't dead but it looked bad. Meyer told him from New York.

"Somebody was going to fix the little bastard sooner or later," Onassis said.[17] He wasn't a hypocrite, and the callousness of his reaction did not surprise Meyer. Onassis told him to call as soon as Kennedy was dead— "as if he wanted to know the result of the four o'clock race at Santa Anita," Meyer later said.[18]

Thring remembers the impassiveness with which Onassis took the news. "Hearing something like that when you are so removed from reality made it even more shocking for me. I didn't expect Ari to be upset; I

---

* Although Jesse Unruh, leader of Kennedy's California campaign, who accompanied the two officers who had arrested Sirhan and were taking him to the Rampart Division of LAPD, originally claimed that this was Sirhan's reply to his question—the same answer Sirhan had given earlier at the scene of the crime, according to an eyewitness, Dr. Marcus McBroom, and reported by United Press and the Associated Press—both police officers denied hearing Sirhan make the statement, and Unruh later claimed he could not remember. But whether Sirhan made the statement or not on June 5, as the political scientist James W. Clarke would later point out (*American Assassins: The Darker Side of Politics*, p. 78. Princeton, N.J.: Princeton University Press, 1982), the record shows that he offered the same explanation repeatedly during the months before and during his trial. Moreover, the explanation is implicit in the notebooks Sirhan kept prior to the assassination. The record, says Clarke, "also reveals that an attempt was made to present Sirhan's preassassination notebooks as evidence of his alleged paranoia rather than as an expression of rational political anxiety, hatred, and preparation for his act, given the assassin's background, values, and perception of Senator Kennedy's position on the Arab-Israeli issue."

knew that Bobby's death was vastly convenient for him; but his reaction was . . . it was as if he'd been told something he already knew."[19]

Rationalizing her remark later, Thring wondered whether the reason for his equanimity at the news was because Jackie had been convinced that Bobby was going to be killed. The night before she left the *Christina,* she had told them that Bobby would die if he didn't pull out of the campaign. "She was quite fatalistic about it. She didn't say he *might* be killed, or he *could* be killed. She said he *would* be killed," Thring recalled.[20] It was more like a prophecy than a premonition.

Onassis and Thring called Jackie in New York.

"Bobby was still hanging in there at that point, but she talked about his death and about Jack's death, and the two seemed to merge in her mind: Bobby became Jack, Jack was Bobby, she was reliving Dallas and crying for Bobby in Los Angeles. She was very distraught. It seemed to her that her country as well as her family was falling apart . . . she was very scared," Thring recalled.[21]

Onassis called Meyer, who was monitoring the news from California at the Pierre Hotel in New York. "He ain't gonna make it, Ari," Meyer told him again. "Call me the minute he don't," Onassis growled.

"I didn't get the impression he wanted to send flowers," Meyer later wryly told friends.[22]

When Bobby's press secretary Frank Mankiewicz announced in California in the early hours of June 6 that Bobby was dead, Onassis called Costa Gratsos. "She's free of the Kennedys. The last link just broke," he said. He still showed no hint of regret, no trace of surprise, merely "a sort of satisfaction that his biggest headache had been eliminated," said a London aide, who had also been called by Onassis that morning.

"Ari had always taken what he wanted, and for the first time in his life he had come up against a young man who was as tough, competitive, and determined as he was. And now that man was dead," Gratsos encapsulated their relationship and the meaning of Bobby's death to Ari when I asked him about it while working on my biography of Onassis.

The night they buried Bobby by candlelight alongside his brother in Arlington Cemetery, Ari said to Meyer: "I guess the kid had everything but the luck." It was the closest anyone heard him come to uttering a word of pity for Kennedy. Nobody was surprised when the man who had been a guest at the White House for the funeral of John F. Kennedy found his name conspicuously absent from the list of those invited to Bobby's funeral service at St. Patrick's Cathedral on June 8. "In the circumstances his presence would have been in very poor taste," David Harlech, one of the ten pallbearers, later remarked.

But perhaps Onassis's strangest and most enigmatic reaction to Bobby's death came in a conversation he had with his sister Artemis a few weeks after the assassination. "I asked him how Jackie was taking Bobby's death. He said, 'She's upset. She misses him; she's blaming everybody.'" Artemis was puzzled when he then compared Jackie's loss to the ticking "death-watch" sound some male spiders make that was supposed to portend death. That sound, he told his sister, was actually to woo female spiders. Jackie, he said, missed that death-watch serenade. If it was his attempt at black humor, Artemis certainly did not get the joke. "He was my brother but he was very complex; I didn't always understand him," she told me.

Meanwhile, on the *Christina*, after they had made love in the final stateroom, Joan Thring asked Onassis how he had amassed his vast fortune. He told her, "There is one thing you must understand about me, my dear: I am completely fucking ruthless." [23]

TWENTY-TWO

A MAN WITH A GRUDGE AND NOTHING TO LOSE

**The mind has great influence
over the body, and maladies
often have their origin there.**

—MICHEL DE MONTAIGNE, 1533–1592

T he small man with tousled black hair who was seized in the melee following the shooting of Robert Kennedy seemed strangely detached. Writer George Plimpton, who had helped in the struggle to disarm him, was taken aback by his "dark brown and enormously peaceful" eyes.[1] Another witness recalled that he looked "very tranquil."[2] His detachment seemed almost transcendental, as if he had an inner life that had no relation to the hysteria around him.

Taken to the Rampart detectives division of the LAPD, where he was booked as "John Doe" and charged with violation of Section 217 of the California Penal Code, assault with intent to commit murder, he carried no means of identification; his fingerprints revealed no criminal record in the state of California. He had four one hundred dollar bills, one five dollar bill, four singles, and change; and he had a cutting of an article titled "Paradoxical Bob" by David Lawrence, clipped from the May 26, 1968, Pasadena *Independent Star-News,* attacking Kennedy for his inconsistency in opposing the Vietnam War while advocating military aid for Israel.

Sergeant William C. Jordan, the night watch commander who had read the assassin his rights and checked the inventory of the objects removed from the pockets of his denim trousers,* was also struck by his curious serenity. "There was more than a touch of mob hysteria in that kitchen after the shooting. It was a very inflammatory situation," said Jordan,[3] who feared that the suspect might be lynched before they could get him back to the station.† Yet, according to the LAPD summary report, the suspect remained less agitated than "individuals arrested for a traffic violation."[4] His demeanor, together with his refusal to talk about the shooting, the one hundred dollar bills, and his apparently calculated anonymity, convinced the police that they were dealing with a "hired killer."[5]

But by midday, the snub-nosed .22-calibre Iver Johnson revolver used in the attack revealed to the LAPD everything their prime suspect would not. Purchased in August 1965 by an elderly man in Alhambra for protection after the Watts riots, it was given to his daughter who, uneasy about keeping a gun in the house around her small children, gave it to a neighbor. The neighbor sold it to a man named Joe, who worked in a department store in Pasadena. Joe turned out to be Palestinian immigrant Munir ("Joe") Sirhan, and he and his brother, Adel Sirhan, identified the prisoner as their brother: Sirhan Bishara Sirhan.

---

* As well as a comb and a car key, there were two unspent .22 cartridges and the copper jacket of a further .22 spent shell, a Kennedy campaign song sheet, and a clipping of a newspaper ad inviting the friends of Robert Kennedy to attend a rally at the Ambassador on June 2.

† Jordan later recommended the two young officers who had accompanied him to the hotel and had protected Sirhan from the mob for a commendation. "They did a hell of a job. If they hadn't done their job just right, we'd still be talking about the fact that another alleged assassin was killed at the scene by an angry mob, and that wouldn't look good," Jordan later told the author.

                                    *   *   *

Sirhan was born in 1944 in a mixed Palestinian and Jewish section of West Jerusalem.[6] When he was four, a few days after the declaration of Israel's independence, his family fled their home to begin a nine-year life of exile in a single room, without furniture, lighted by a single kerosene lamp, in the Old Walled section of the city.[7] He was twelve when, in 1956, with the help of Lutheran missionaries and the United Nations Relief and Works Agency, his family moved to Pasadena, California. But before the year ended, their father returned to Jordan, which added "a sense of abandonment to the feelings of isolation the family already was experiencing in a totally new and alien environment,"[8] wrote James Clarke in his excellent book, *American Assassins*. They still regarded Palestine as their home; their friends, their interests were all in Palestine. They continued to speak Arabic, listen to Arabic music, read Arabic newspapers, and observe Arabic customs—all in the hope that someday they would return to their homeland.[9]

Nevertheless, Sirhan appeared to have adjusted well to life in America. In 1963, he graduated from the John Muir High School ranked 558 in a class of 829.[10] But beneath the surface, the violence and injustices of his past were catching up with him. In a school history book, he had underlined a passage describing the assassination of President McKinley—"After a week of patient suffering the President died, the third victim of an assassin's bullet since the Civil War"—and added in the margin: *"Many more will come."*[11]

He became well informed on Middle East affairs, an interest to which his library card attested.[12] His future, he believed, was there—not in the United States.[13] He attended meetings of the Organization of Arab Students, whose members recalled his fierce Arab nationalism and his hatred of Zionists, whom he considered to be worse than Nazis.[14]

Expelled after only four semesters at Pasadena City College,[15] he got a job as a stable boy at the Santa Anita race track, and had ambitions and the frame—he was five-feet-five, weighed about 120 pounds—to become a jockey. The following summer, he became an exercise boy at the Graja Vista Del Rio Ranch in Corona, California. But a fall at full gallop on a misty morning in 1966 left him with blurred vision that put paid to his aspirations. Not long after this, the Six-Day War brought home to him the fact that his dream of one day returning to his own country was probably over, too. He expressed his frustrations in increasingly fierce political arguments. But even those friends who had seen him work himself into a rage over the Palestine issue, and America's support for the Jews, were shocked when he murdered Bobby Kennedy.

The crime had been committed in public; Sirhan's culpability was never an issue, and within hours of the shooting Los Angeles mayor Sam Yorty disclosed to the press Sirhan's identity, his Palestinian roots, and the fact that detectives had found incriminating notebooks containing threats to Kennedy's life in his room at his family home, "It appears," Yorty told reporters, "Sirhan Sirhan was a sort of loner [who had] indicated that RFK must be assassinated before June 5, 1968. It was a May 18 notation in a ringed notebook."

Yorty's remarkable revelation still gave no indication of the sensational content of Sirhan's notebooks, one page of which read:

> *May 18 9:45 AM—68*
> *My determination to eliminate R.F.K. is becoming*
> *more the more of an unshakable obsession . . .*
> *R.F.K. must die—R.F.K. must be killed Robert*
> *F. Kennedy must be assassinated R.F.K. must*
> *be assassinated R.F.K. must be assassinated . . .*
> *Robert F. Kennedy must be assassinated before*
> *5 June 68 Robert F. Kennedy must be assassinated*
> *I have never heard please pay to the order of of of*
> *of of of of of of of this or that please pay to*
> *the order of*

The police had found three notebooks containing similar tortured streams of consciousness scrawled across their 9-by-12-inch pages. In the words of American investigative reporter Bob Kaiser, they were "a detective's dream . . . if that didn't prove premeditation nothing did." [16]

*   *   *

"When I read that they'd arrested a Palestinian my first thought was, 'Well, Mahmoud's done it'; my very first thought was, 'Hamshari's finally done it," said Mohammed Ibrahim, who had been at the secret crisis meeting in Damascus after the humiliation of the Six Day War, where Hamshari proposed targeting a high-profile American in America. A former Fatah intelligence agent close to Khaled Hassan's conservative faction, Ibrahim told me in Switzerland in 1999, "I got mad [when Bobby Kennedy was shot] because I felt we could not have chosen a surer way to discredit our cause."[17] But to his surprise, the announcement from Fatah claiming responsibility for the killing never came. "I knew that Hamshari had been sent to Los Angeles for a big job; it was inconceivable to me that he was in Los Angeles and had not had a hand in Robert Kennedy's murder," said Ibrahim. Obviously, Sirhan had pulled the trigger, he said, "but political assassination is seldom the work of one man. Organizations, not individuals, are usually behind killings of that importance. A lot of money and preparation had gone into setting up Hamshari [in Los Angeles and Paris]. It would have been ironic if somebody else had got in ahead of him [to kill a high-profile American in America]."

Sirhan's Palestinian roots quickly became an issue, and suspicions alighted on Fatah. In the editorial view of *Life* magazine, Sirhan "seemed formed in the classic mold of political assassin—small, proud, polite, repressed and aboil with a secret, almost religious sense of cause: Arab nationalism."

A reporter and feature writer for *The Arizona Republic,* and Rome correspondent for *Time* magazine, Bob Kaiser served as investigator for Sirhan's defense team. Later, in his seminal book of the trial, *R.F.K. Must Die!,* Kaiser would claim that the police were less than diligent in following up on Sirhan's alleged links with Fatah. "Sirhan had Arab friends, for example, who had left the United States and returned to the Middle East for no apparent reason, but neither the police nor the FBI had enough interest in them to interview them there," he wrote. One young man seen in still photographs of Robert Kennedy's last talk on the Embassy Room stage was recognized by a Kennedy campaign worker as the same man who had visited Kennedy headquarters in mid-May; identifying himself as Ali Ahmand, an employee at Microdot in Pasadena. By the time detectives got around to Microdot three months later, reported Kaiser, fellow workers said the man had quit the company and gone back to his homeland of Pakistan. And his name wasn't Ahmand at all, but Iqbal. "Checking police reports and FBI records of the investigation on file in the L.A. district attorney's office, Kaiser discovered that "neither the police nor the FBI had bothered to check on Iqbal in Pakistan, ask him why he had given a false name to the Kennedy worker in mid-May, or what he was doing on the Embassy Room stage."[18] (When I asked Ibrahim what he made of this lapse in the investigation, he told me: "He could have been Fatah, or some innocent Pakistani fellow, or he could have been Hamshari himself!")

Yet despite the rumors of a Fatah connection, there was still no announcement from the terrorist organization in Jordan. Ibrahim called it "a case of collective political amnesia."

In fact, there are three possible explanations for the silence. First, Arafat's popularity had risen dramatically following the battle of Karameh. He had become "Mr. Palestine," the embodiment of his country's aspirations; and the leadership of the PLO *—the central body of all the Palestinian resistance groups—was suddenly within his grasp. The position would mean his ascension from terrorist leader to a politician of importance and a player on the world stage. With so much money, power, and prestige at stake, it was not a good moment for Fatah to claim responsibility for the murder of a man whom many believed would have been the next president of the United States and leader of the free world.

Second, only a handful of the most loyal and trusted people in Fatah had heard the assassination proposal Hamshari made in a small hotel room in the desperate aftermath of the Six-Day War, and that appeared to have been rejected in favor of his more conciliatory idea of seeking American support. Even fewer knew that he had indeed gone to Los Angeles a few weeks later, and might not have made the connection at all.

Third, although the original idea might have been Hamshari's, the operation—whether it was to assassinate a prominent American, or simply

---

* Palestinian Liberation Organization.

to raise funds—was almost certainly controlled by Fatah's intelligence chief, Abu Iyad. "The fixer, the man for confidential missions, the keeper of PLO secrets," the authoritative PLO historian Patrick Seale calls him, and he and Fatah's military chief Abu Jihad "ran their own autonomous outfits with their own loyalists, much as barons might do under a medieval king." [19] Therefore, while it might seem inconceivable that an assassination of such significance could have been contemplated without Arafat's being at least aware of it, even if he had remained fastidiously ignorant of specifics, it was possible that Abu Iyad had, in order to preserve the integrity of the man who was the closest person Palestine had to a head of state, deliberately kept him in the dark. And to let him remain in the dark after Kennedy's murder suddenly became a potential embarrassment to his—and Fatah's—hopes of taking control of the influential PLO.

But the fact remains that one way or another—perhaps because nobody knew the truth, and perhaps because a few feared the worst—Fatah closed ranks. (According to Mohammed Ibrahim, when he attempted to voice his own doubts with a prominent Fatah dove shortly after the assassination, he was told to forget it—unless he wanted to be "sewn in a blanket and buried in the Badiyat ash Sham [Syrian desert]." *) [20]

*   *   *

---

* I was introduced to Mohammed Ibrahim by Yannis Georgakis at the Hotel Grande Bretagne in Athens in 1992. Georgakis had been Greece's ambassador-at-large to the Arab oil-producing countries, and had an impressive circle of interesting, important, and shady contacts in the Middle East. Ibrahim was one of many members who had parted company with Fatah in questionable circumstances. "I believe there was some disagreement over missing funds," was all that Georgakis would tell me about the incident, although he vouched for the Palestinian's bona fides, if not his name, which Ibrahim sometimes switched to Hassan. At the end of our first meeting, Ibrahim agreed to see me again in Paris in one month's time; he would not give me a contact number, nor an address; at the second meeting he told me that if I wanted to contact him again, I must take an advertisement in the personal column of the *London Times,* saying: "Camille. Pour l'amour du Grec. Peter;" and he would get in touch with me. I took my first ad on May 18, 1995; he called me the following week and we arranged another meeting. He answered my questions, gave me other leads, and disappeared. I did not talk to him again until Friday, September 21, 2001. He called me from the Savoy hotel in London and suggested we meet that evening.

It is, of course, the nature of conspiracies that we are never told the whole truth. And if, as Patrick Seale claims, it is characteristic of the hothouse of Palestinian politics that "every man's hand is raised against his brother,"[21] what is to be made of Ibrahim's story? A former Fatah dove, who, in his own words, "quit the reservation" in the 1970s, is he simply a man with an old score to settle?

Possibly. Nevertheless, his story also resonates with a piece of information from a separate Fatah source, who claims to have given Hamshari, in the autumn or late winter of 1967, a list of Palestinian immigrants living in Los Angeles. The list was acquired from the United Nations Relief and Works Agency, responsible for the welfare of refugees in America—and one of the organizations that Sirhan's mother, Mary, had approached for financial help to emigrate with her family to America twelve years earlier.

Although things are not always as they seem, this much is sure: If Hamshari had gone to Los Angeles to find an assassin (rather than raise funds), he must have known that he had found his man the moment he met Sirhan.* With his anti-American chip and Palestinian fervor, Sirhan was a man with a grudge and nothing to lose.

It was not possible, I told him. My younger sister had died, and I was leaving that after noon for her funeral on the Greek island of Corfu, where she had lived for several years. Ibrahim sounded agitated when I told him this; he had some "important intelligence" for me and implied that I was being unreasonable in refusing to cancel my journey and meet him that evening. Eventually, he suggested that I break my flight back to London to meet him at the Grande Bretagne for lunch the following Monday. I agreed. He stood me up. When I had heard no more from him by November, I placed another ad in the *Times:* "Hassan. The first blow does not fell the tree. Urgent you call. PE." It was a phrase he had often used when he thought I was being too impatient or wanted to avoid answering my questions. He called, apologized, and said that he would call again to fix an appointment as soon as he could sort out his "desperate, sad life." He repeated that he had "important intelligence" for my book. But a few days later, on November 28, I had a visit at my London home from two officers of the British antiterrorist branch at Scotland Yard. From their questions, it was evident that they were as keen to talk to Mohammed Ibrahim as I was. I have not heard from him since.

* It is unlikely that Sirhan was sought out as a potential contributor. Out of work, the most he could afford could only have been a modest contribution from the $1,705 set-

*   *   *

Onassis's trust in Mahmoud Hamshari's abilities was deeply shaken
when the LAPD established a Palestinian connection so fast. Suddenly
he was "a rich guy on a yacht with too much time on his hands," Meyer
told Brian Wells about how Onassis began bombarding him with questions: Was Sirhan a passing lunatic? A hired killer? What was his story? "I
said, 'Who knows why anyone kills anyone any more, Ari? Last month it
was Martin Luther King. This month it's Bobby Kennedy. Next month
it's gonna be some other poor bastard,' " Meyer recalled his reply.[22]

Bobby died in the early hours of June 6.

In Athens, a small cloud of apprehension was beginning to gather in
Georgakis's mind about the size of the Hamshari settlement. He knew he
had enemies at the airline, and began to worry about leaving himself
open to accusations of profligacy (which, indeed, would cause his downfall the following year*). He called Gratsos in New York and, under the
pretense of discussing his concern about Onassis's health, and the pres-
sures on him because of the problems with the colonels, he mentioned
the amount of the Hamshari payment—$1.2 million, including the
$200,000 which he himself had provided in cash from the Olympic
coffers—as an example of Onassis's questionable judgment.

What Gratsos really made of this figure—more than three times the
sum Hamshari had asked for at their original meeting at avenue Foch—
is unknown. But according to Georgakis, Gratsos told him to forget it.
The deal was done, the airline was safe, and that was all that mattered.
Blackmail is not a situation that allows much room for hard bargaining,
he told Georgakis philosophically.

---

tlement of a workmen's compensation claim he had recently received from the Ar-
gonaut Insurance Company for his fall from the horse at the Corona racing ranch. (Trial
testimony, Vol. 19, pp. 5427–5445.)

* According to Miltiadis Yiannakopoulos, "Georgakis was very good but he had no
notion of money. Once when he was ill, he flew to New York to get treatment. On his
bill he had a tip for $4,000. That enraged Onassis. He let him go, although he kept him
as an adviser."

TWENTY-THREE

A MARRIAGE OF SALE

**I remember thinking to myself,
'This is a marriage of sale.
Where is the love?'**

—CONSTANTINE HARITAKIS, ON THE MARRIAGE
BETWEEN JACKIE AND ONASSIS

I hate this country. I despise America, and I don't want my children to live here anymore," Jackie exploded after Bobby Kennedy's funeral. "If they are killing Kennedys, my kids are the number one targets . . . I want to get out of this country and away from it all!"[1]

She had always believed that nothing in life was more precious than privacy, and no one was better able to give that to her than Onassis: aboard his yacht, on his private island, inside the guarded splendors of his homes in Paris and Athens, or in one of the permanent suites he maintained in the finest hotels from London to Buenos Aires: Only the very rich, she knew, owned houses to conceal their whereabouts. Now, with Bobby gone, there was no reason to delay any longer, and two weeks after the funeral, she called her mother and asked if she could bring a house guest to Hammersmith Farm for the weekend.

"Mummy said, 'Yes. Who is it?' 'Aristotle Onassis,' said Jackie. 'God, no. Jackie, you can't mean it,' " her half-brother Jamie Auchincloss would later recall of his mother's undisguised repugnance at the idea. " 'You just can't mean it.' "[2] But Jackie meant every word of it, and all summer long—shuttling between Cape Cod, Newport, and Palm Beach—she and Onassis pursued a charm offensive to win over the people who mattered most in a world that still meant so much to her social heart.

Out of loyalty to Jackie, Onassis was tolerated; but there was no real softening toward him; his affair with Callas, his Levantine looks, and the suspicious complexity of his past were not the kind of credentials that got a chap sponsored for the Clambake Club. "He told a good story, but he seemed to expect applause, as if he were an entertainer," one Cape habitué told me.

Onassis saw through their *politesse*. "They hate my Greek guts," he told Gratsos.[3] He believed this not out of a sense of persecution, but out of experience: Their world would never be his world—he bestrode it like a *maitre d'*, sniped one of Jackie's gay friends. Nevertheless, their subtle ostracism was a source of amusement as well as anger to Onassis. "Everybody here knows three things about Aristotle Onassis," he answered when Johnny Meyer asked how he was coping with Jackie's crowd: "I'm fucking Maria Callas. I'm fucking Jacqueline Kennedy. And I'm fucking rich."[4]

There were, however, occasional flashes of his irritation at "being led around by Jackie like a prize bull."[5] When one Palm Beach matron asked what he thought made him so attractive to women, he told her: "Frankly, my dear, I'm better hung than most guys."[6] The painter Rico Zermeno, who had known Jackie since her first year at the Sorbonne in Paris in 1949, and who became one of Onassis's favorite clubbing companions in the South of France, had no doubts about the veracity of this statement: "You ever see the picture of Jackie and Ari naked? Ari had a huge cock, and ladies I've talked to say he knew how to use it, too."[7]

Nevertheless, as Onassis continued to be paraded around from one select group of Jackie's smart friends to another, *his* friends were wondering why he went along with it. The reason for their surprise and confusion was that a few weeks before Bobby's death, Onassis started to change the story about his feelings toward Jackie that he was telling his American friends: He did not want to marry Jackie; he wanted to marry Maria Callas; it was Bobby who was forcing him into a shotgun wedding with Jackie!

Emmet Whitlock, the former head of the International Petroleum Freight Exchange in New York, and a man well connected in the Palm Beach social establishment, said that he completely understood Onassis's dilemma; he also understood why Bobby had issued the ultimatum.

"Onassis and Jacqueline Kennedy had been sleeping together for quite some time," he told me. "Their affair had been going on for years. My wife and I were asked down to Nassau and Gloria refused to go because Mrs. Kennedy was going to be Ari's girlfriend on the boat, and she felt that that was rather a dirty trick to play on Maria. That was in 1964. Jacqueline and Ari were having an affair then. Yes, they were. In 1964, less than a year after Dallas, they were already having an affair. Of course, it was all hush-hush. But Bobby was worried about it. [As head of the preeminent Catholic family in America, and about to run for the highest office in the land], he didn't want the truth about his sister-in-law's affair with Ari, and how long it had been going on, to come out in the run-up to the election. That wouldn't have done at all. So he demanded that Jacqueline and Ari get married. Ari told me that himself.

"Now what possible hold could Bobby Kennedy have over Onassis to tell him he can't marry Maria—and must marry Jackie?" Whitlock, a slim, handsome man in his eighties, went on. The answer, he believed, was this: "Onassis had decided that the future of the tanker business was in 100,000-ton tankers. Now 100,000-ton tankers draw about 90 feet of water; they are the cheapest way to move oil, but they can move it to very few places in the world because of their draught. Now Ari operated very cleverly, and had set up his children as U.S. citizens, and used them to get government loans to build his tankers through the U.S. Maritime Commission.

"So now Bobby Kennedy says to him, 'You are going to marry Jackie.' Ari says, 'Oh no I'm not.' Bobby says, 'Oh yes you are, because if you don't I'll pull the plug on you—on your loans, on your access to U.S. deepwater ports.' Ari had seven million dollars of his own money tied up in this thing. And that's how Bobby Kennedy made Ari Onassis marry Jackie Kennedy. Absolutely true."[8]

But why, then, had Onassis continued to jump through Jackie's hoops in Cape Cod, Newport, and Palm Beach after Bobby Kennedy's death? According to another prominent Palm Beach resident, William Carter, former Vatican ambassador to the United Nations, knight of Malta, and long-time Onassis intimate—to whom Onassis had told the same story about Bobby putting a gun to his head to force him to make a respectable woman of Jackie before the presidential election—it was possible that the gun had passed to Teddy Kennedy.

However, that presupposes that a gun existed in the first place, Carter mused skeptically. "I never bought Ari's story that he didn't want to marry Jackie. He was desperate to marry Jackie; it was business at first sight. He must have had his own very good reasons for lying about it. It was very strange, but the molecules of truth were buried very deep in Ari's psyche."[9]

By this time a kind of momentum had built up, and with Bobby barely cold in his grave, Sirhan still awaiting trial in Los Angeles, and the Palestinian connection exposed, Onassis had good cause to be wary—which is why his story of Bobby's enthusiasm for the marriage was so important. For it at least made Bobby's death seem a little less convenient as Jackie proceeded clearing the path to her wedding to the man he had actually vowed she would marry over his dead body.

Not a single friend thought Jackie should marry Onassis. But now that Bobby was gone, there was no one who could stop her. "She could completely seal off anything unpleasant almost the moment it happened to her. It was part of what made her appear so untouchable and, it must be said, rather selfish sometimes," said Billy Keating, an American art dealer who had known her since 1956, when she fled to London after walking out on John Kennedy.[10]

*   *   *

On July 28, Jackie spent her thirty-ninth birthday at Hyannisport. Matriarchal Rose organized a family dinner, to which Onassis was not invited. That evening Jackie told Ted Kennedy, the last surviving brother and effectively the head of the family, that she was going to marry "the Greek" in the autumn.

Ted had taken Bobby's murder even harder than he had Jack's, and, drinking heavily, he was really in no condition to deal with Jackie's problems on top of his own. But even in his most befuddled alcoholic state, he must have known that she would never be as loyal to him as she had been to Bobby, and that there was an element of revenge in her decision to go off with a man who had cuckolded Jack and whom Bobby had died hating. Kennedy friend George Smathers said that going off with Onassis was Jackie's way of humbling the Kennedy women, who had always "flaunted their money and power." It was her chance to say to them: " 'Okay, what are you going to say now that I can buy and sell you?' "[11]

Yet some Kennedy intervention in the matter was still required, and in August Ted called Onassis and asked for a meeting. Onassis invited him to Skorpios.

Ted was no problem for Onassis, who would later liken his role to "a priestly hustler peddling indulgences."[12] He listened patiently while Kennedy reminded him that to millions of Americans Jackie was an icon; that it would cause great indignation if a foreigner were to become the stepfather of the late president's children. Ted doggedly played out the game: religion, the children, politics, family loyalties. But however much he dressed it up, in Joan Thring's words, "Teddy was negotiating for Jackie's body."[13]

"We love Jackie," Ted eventually concluded his case.

"So do I, and I want her to have a secure life and a happy one," Onassis replied.[14]

But she would automatically forfeit the $150,000 a year she got from the Kennedy trust if she remarried, Kennedy said; even her presidential widow's pension would cease.

They had gotten to the nub of the issue.

Both men understood that the shame of the market is not in being sold, but in being undersold. According to Georgakis, Onassis offered three million dollars for Jackie, plus one million dollars for each of her children; he would be responsible for her expenses so long as the marriage lasted; after his death she would receive $150,000 a year for life, exactly the amount she would have received from the Kennedy trust.

The Athens jeweler, Costas Haritakis, was in the room when Onassis wrote out the agreement in longhand. "I remember thinking to myself, *'This is a marriage of sale. Where is the love?'* Ari never showed any sign of excitement for the marriage. He was simply doing business."[15]

Nevertheless, Jackie's friends were far from pleased with the deal Ted Kennedy brought back from Skorpios. Ted was not naive, nor was he without his own kind of smarts, but he seemed to have been completely outplayed by Onassis. "Ted Kennedy couldn't negotiate buying a pack of condoms without the help of his aides," sniffed the socialite-writer Taki Theodoracopulos,[16] and certainly Jackie did not hide her disappointment at the mediocrity of the settlement he produced. "She felt that Ted Kennedy had been blown over by the force of Ari's personality—and that he might also have hit the bottle a little too hard on the island," said Nigel Neilson.[17]

But whatever Kennedy's clinical condition, were his hands tied even before Onassis started pouring the ouzo at the party he gave on the eve of their talks?* According to Georgakis, Onassis went into the talks on

---

* A member of the bouzouki band brought from Athens for the occasion was Greek journalist Nicos Mastorakis, whose notes give some idea of the mood of the evening: "Teddy holds a blonde goddess . . . Teddy drinks ouzo, permanently. Jackie prefers vodka at first . . . the bouzouki music reaches its peak and Teddy gets up and tries to dance . . . Teddy returns to his ouzo." Eventually Mastorakis was seen taking pictures and the film was removed from his camera. "Kennedy told him, 'If you print one word out of place about this, if you hurt me, I'll have your ass," Johnny Meyer recalled. "I don't know what he thought he was going to do, the senator from Massachusetts, to this Greek kid. But he was right to be worried. He'd had a snootful, he hadn't exactly been a boiled shirt all evening." As I revealed in *Ari*, Onassis used his pull with the colonels, and Mastorakis was picked up in Athens. "I guess they put the frighteners on him because he did an okay piece, although later Ari showed me a copy of his original story, which had been cut to ribbons by the military censors. If it hadn't been for Jackie, Ari wouldn't have given a damn. He didn't have much time for Teddy," said Meyer.

Skorpios perfectly relaxed not because he believed that Ted Kennedy was no match for him, but because he knew that when it came to the crunch he held the high cards. These, he hinted to Georgakis and Nigel Neilson one evening, were the fact that he had been Jackie's lover *before* John Kennedy died—and that after Dallas, he had *shared her with Bobby.*

"It was often difficult to judge how *rational* Ari was sometimes," Neilson later told me. "He was capable of playing that card, because it was in his hand, and because he loathed the Kennedys so much, but whether he actually played it, I have no idea. Only two men were in the room [when Jackie's prenuptial settlement was discussed] and that was Ari and Ted Kennedy, and Ari never talked about it afterward, and I can't imagine Kennedy ever will." [18]

But plainly Kennedy would not have underestimated Onassis's determination, and any suggestion of the story getting out would clearly have unsettled his resolve to strike a hard bargain on Skorpios. For although he had declined to fight for the Democratic presidential nomination that year—this is "Bobby's year," he had told aides who believed that it was his for the asking[19]—Ted's time would surely come, and the memory of Jack and Jackie, and the romance of Camelot, would be as important to him as it had been to Bobby. To have the legend sullied by a sordid sex scandal would have been a terrible waste.

Even though Onassis had not expected Jackie to accept his first offer relayed to her by Ted Kennedy, he was appalled at the "naked greed" of the counterproposal—$20 million, up front—put forward by her financial adviser, André Meyer, head of the investment banking firm of Lazard Frères (André "did not feel that it was a good merger," gibed an associate).[20]

"Your client could price herself right out of the market," Onassis said roughly when he confronted Meyer on the evening of September 25, at the banker's apartment in the Carlyle Hotel in New York.[21] The remark began a long and seriously unpleasant session in which Meyer repeatedly justified his reputation as a man who would fight tooth and nail for the ultimate buck. The deal Meyer was said to have cut for Jackie was impressive, indeed; but whatever the truth of it was—one widely reported figure at the time claimed that Jackie would receive $100 million if Onassis died while they were still married—the contract clearly did not stand up, as we shall see.

On October 15, 1968, the *Boston Herald-Traveler* elevated the rumors that had been growing since Kennedy returned from Skorpios to a page one story: JOHN F. KENNEDY'S WIDOW AND ARISTOTLE ONASSIS TO WED SOON. Kennedy son-in-law Steve Smith called Pierre Salinger in Washington. "We have to figure out some kind of statement for the family to put out," he said.

"Have you got any idea of what you want to say?" Salinger asked.

"How about," said Smith, who had majored in philosophy at Georgetown, " 'Oh shit!' "[22]

Since Ted Kennedy was adamant that the wedding should not take place in the United States, and Jackie had expressed a wish to be married on American soil, Georgakis was dispatched to the U.S. embassy in Athens to ask for permission to hold the ceremony there. The ambassador passed when he heard the names of the couple involved. In desperation, Georgakis suggested the chapel on Skorpios. Nobody expected Jackie to agree to this, but she did. Her only proviso was that they find a priest who "understands English and doesn't look like Rasputin."[23]

Christina wept inconsolably when she heard the news. Alexander, who disapproved of Jackie as strongly as his father disapproved of Fiona, vowed he would not go to the wedding.*

Reached in Paris, Maria Callas told newsmen enigmatically, "I'm happy for those who are happy."[24]

Lee Radziwill who, according to Truman Capote, still "kind of had it in mind that she was going to marry Mr. Onassis herself,"[25] said she was "very happy to have been at the origin of this marriage, which will, I am

---

* Fiona eventually persuaded him to attend as a mark of respect to his father. "Well, it's certainly a perfect match," he told her wryly. "The old man loves names and Jackie loves money."

certain, bring my sister the happiness she deserves." Privately, it was another, sadder story. *"How could she do this to me!",* she screamed at Capote on the phone when she heard the news. *"How could she! How could this happen!"* [26]

Ted Kennedy's statement wishing the couple well was "chilling in its formality and brevity," noted *Time* magazine.

There was a terrible storm over Skorpios on October 19, as the guests gathered on the eve of the wedding. Jackie was accompanied by her children, her mother and stepfather Hugh D. Auchincloss, and Lee. Rose Kennedy refused to go, and sent Pat Lawford and Jean Smith in her place. It was not a joyful gathering. Apparently still wondering how Jackie could have done this to her, Lee appeared to have decided to forget the import of what she had lost in the good champagne.

Later that night she would find other ways to alleviate her sorrow.

It was still raining on Skorpios on October 20, but now it was that fine rain that comes and goes in the autumn on the islands in the Ionian Sea. It was a good omen, insisted Onassis's sister Artemis, though privately she was angry with her brother for making love to Jackie in the afternoon before the ceremony. Certainly it was clear that Jackie had not been resting all afternoon when she came on deck of the *Christina* shortly before the ceremony at 5:15 P.M. "I noticed that the last three buttons of her dress were not properly fastened; they were in the wrong loopholes," Onassis's best man, Costas Haritakis, recalled. "She wore a Valentino dress which had tiny pearl buttons from the top to the waistline. It was a beautiful dress but it was crumpled as if she had made love in it. Was this the importance she gave to being married to this extraordinary man—she couldn't even be bothered to change her dress?" [27] To see your bride before the ceremony on the day of the wedding is considered to be bad luck; to make love to her is tempting fate, Artemis told her brother angrily.

Around the world front pages were filled with damning headlines: *Jack Kennedy Dies Today for a Second Time,* reported Rome's *Il Messagero.*

The Reaction Here is Anger, Shock and Dismay, proclaimed the *New York Times.* When Vatican sources suggested that having violated the law of the church, Jackie was ineligible to receive the sacraments, her spiritual adviser, Boston's Cardinal Cushing, now practicing damage control for the Kennedys, threatened to resign.[28]

But Jackie had plenty to assuage her hurt.

Guests were agape when she arrived at the reception aboard the *Christina.*[29] On her left hand sparkled a huge heart-shaped ruby surrounded by diamonds; from her earlobes hung two equally immense rubies, also set in diamonds. Johnny Meyer said that she looked like a million dollars. Haritakis, who had sold Onassis some of the stones, knew that it was closer to one million-five.

Only Onassis knew the true cost.

Two days later, Jackie made it clear what kind of wife she intended to be: She would not accompany Onassis to Athens to meet Prime Minister George Papadopoulos, nor would she be present when they announced Project Omega. Her presence would have been a public relations coup that the *junta* desperately needed, and Onassis was furious. He was asking her to do for him no more than she had done for John Kennedy, he reminded her testily.

This, of course, was true, although it is one thing to be used by your husband when he is the President of the United States, and you are the First Lady; it is another matter altogether when he is a wheeler-dealer on the make, and you are just another trophy wife. "I've made some terrible mistakes in my life, but marrying Jackie might take the biscuit," he told Georgakis when, forty-eight hours after their wedding, he turned up in Athens without her. When Georgakis suggested that his bride might die of boredom on Skorpios without him, Onassis told him furiously: "She's busy writing to all her old lovers."* Georgakis wondered whether Jackie had any idea that he was "already intercepting her mail and tapping her

---

* "Dearest Ros," she wrote to Roswell Gilpatric, "I would have told you before I left— but then everything happened so much more quickly than I'd planned. I saw somewhere what you had said and I was touched—dear Ros—I hope you know all you were

phones?[30] (Eventually, Jackie did find out; although how is unclear. "Opening my mail! Who did he think he was? Edgar Hoover?" she told her friend, Billy Keating.[31])

Meanwhile, stung by Jackie's failure to attend the Omega announcement party, the colonels began to have doubts about the rest of the deal—especially since Niarchos had stepped in with an even better offer, although one that was contingent on his getting the oil refinery which was at the heart of Onassis's deal.

Bribed up to his epaulets by Onassis, confronted by colonels who had been similarly seduced by Niarchos's largesse, Papadopoulos did what all politicians do when they are in a jam: he ordered an inquiry into the two bids.

Five months had gone by since Bobby's murder. Neither the LAPD nor the FBI had uncovered evidence of a conspiracy; no alarms had gone off at the CIA; and even Sirhan's own lawyers—planning to plead diminished capacity and lack of premeditation, an awkward argument to sustain if their client had been involved with others—were content to go along with the LAPD conclusion that there had been no conspiracy. Unfortunately for Sirhan, this line also enabled the prosecution to concentrate on their main goal of sending him to the gas chamber without—as Professor Philip Melanson, director of the Robert Kennedy Assassination Archives* has pointed out—having to "worry about the legal complexities of a conspiracy."[32]

Onassis must have been breathing easier.

On October 31, eleven days after the nuptials on Skorpios, and the night he was reported dining with Maria Callas at Maxim's while Jackie continued to languish on Skorpios, he called Georgakis and told him

---

and are and will ever be to me—with my love, Jackie." Although it was barely more than the kind of note that is regarded in Jackie's milieu as good manners, it infuriated Onassis; and Gilpatric's wife sued for divorce the day after its publication in 1970 when it fell into the hands of an autograph dealer in New York.

* Southeastern Massachusetts University.

that he had changed his mind about the Hamshari payments. He would not pay another penny, he said. This was the first time he had mentioned Hamshari to Georgakis since he had asked for the initial $200,000 to take to New York, and Georgakis was surprised that the question had come up again, since he believed that the matter had been settled.

Nevertheless, it is not difficult to imagine what had caused Onassis's change of heart. The idea that Bobby was going to be killed had been around even before he announced his decision to run for the presidency; Jackie was convinced of it; Bobby himself had a premonition about it; even the reporters covering his campaign felt that the killer was "out there now waiting for him."[33] Had Hamshari, therefore, simply gambled on the fact that if Bobby were to be murdered—*by anyone*—Onassis would assume that it was the fulfilment of their contract?

When no evidence of a conspiracy was found in Los Angeles, and even Sirhan's lawyers could only resort to a plea of diminished capacity and lack of premeditation, this explanation must have been irresistible to Onassis's own Machiavellian mind.

By this time, however, the Palestinians had hijacked their first Boeing 707; intelligence agencies had warned airlines to expect more attacks; and when Georgakis demurred about double-crossing Hamshari, Onassis replied that none of the other airlines had paid nearly as much as they had, and Hamshari had gotten all he was going to get!

On the night of December 19, Georgakis was dining with Onassis at his Glyfada villa when Onassis got a call from his secretary in Paris: A man had called on Onassis's private line and said that a bomb had been placed aboard the evening Olympic flight out of Kennedy International Airport to Athens—the flight on which Jackie, her two children, and four friends were booked to fly to Greece for the Christmas holidays.

The aircraft was turned back just as it was taxiing toward the takeoff runway. The flight was delayed more than four hours while the NYPD bomb squad checked the plane and every piece of luggage.[34] No bomb was found. Nevertheless, since Jackie's booking had been a closely guarded secret—until she arrived in the VIP lounge, only her own Secret Service people and the Olympic operations manager knew that she would be flying out of New York that evening—the incident exposed a serious breach of security.

Early the following morning, Georgakis got a call from Hamshari. Unless the money Onassis owed him was paid into his Swiss account by three o'clock that day, he said, the next threat would not be a hoax. Georgakis called Onassis and told him that all Olympic flights would be grounded until he settled with Hamshari. Clearly shaken, Onassis at last agreed to pay up.

A few days later, an El Al Boeing 707 waiting to take off from Athens for New York was attacked by two Palestinians; one person was killed, and several others were seriously wounded.* Georgakis called Onassis. "We came that close, Ari," he told him, still angry at how casually he appeared to have treated the threat.[35]

---

* On July 27, 1970, the two terrorists arrested and sentenced to life for this attack (December 26, 1968) were released when an Olympic Airways plane flying to Cairo was hijacked by six Palestinians. (E. F. Mickolus, *Transnational Terrorism: Chronicle of Events 1968–79*. Westport, Conn: Greenwood Press, 1980; and Christopher Dobson and Ronald Payne, *War Without End: The Terrorists: An Intelligence Dossier*, p. 242. London: Harrap, 1986.)

TWENTY-FOUR

SIRHAN'S WHITE FOG

Wars begin when you will, but
they do not end when you please.

—NICCOLO MACHIAVELLI, 1469–1527

On February 12, 1969, after six weeks of preliminaries and jury selection, Sirhan Bishara Sirhan stood trial in the Los Angeles County Superior Court under accusations of having killed Senator Robert F. Kennedy. There was no mystery about who fired the gun: Kennedy was shot by Sirhan in a roomful of witnesses, and the prosecution depicted the Palestinian as a nerveless plotter who had hunted Kennedy down and, "alone and not in concert with anyone else," killed him in cold blood.

Opening for the defense, Emile Zola Berman pictured his client as an "immature, emotionally disturbed and mentally-ill youth," intellectually incapable of premeditation, and "in a trance" when he killed Kennedy.[1] Sirhan's boast that he had shot Kennedy because he was a friend of the Zionists, and might possibly have been the next president of the United States, was ignored by his lawyers. Despite his anger at being portrayed as mentally ill, Sirhan knew that this was his best line of defense. If he could avoid the death penalty he believed that he might eventually be repatriated to a friendly Arab state in a prisoner exchange deal; as he whispered to one of his defense team in court, "Better a live dog than a dead lion."[2]

Testifying for the defense, University of California psychiatrist Dr. Bernard L. Diamond diagnosed Sirhan as being "out of control of his own consciousness and his own actions, [and] subject to bizarre dissociated trances in some of which he programmed himself to be the instrument of assassination."[3]

Meanwhile, Special Unit Senator, the LAPD's assassination task force, had constructed a meticulous timetable of Sirhan's activities prior to the shooting. "We took him back for more than a year with some intensity—where he'd been, what he'd been doing, who he'd been seeing. But there was this ten- or twelve-week gap, like a blanket of white fog, we could never penetrate, and which Sirhan himself appeared to have a complete amnesia block about," says Bill Jordan, the night watch commander at Rampart detectives who was Sirhan's first interrogator.[4]

After Sirhan emerged from this "white fog" in March 1968, he joined the Ancient Mystical Order of the Rosae Crucis, a San Jose organization offering salvation through mysticism and mind control[5]; he also began to frequent bookshops that specialized in the occult, and started scrawling trance-like entries in his notebooks, which defense psychiatrists identified as "automatic writing"—a technique sometimes used by hypnotherapists to extract information from, or implant ideas in, the subconscious of a hypnotized patient:

> *May 18 9:45 AM—68*
> *My determination to eliminate RFK is becoming*
> *more the more of of an unshakable obsession . . .*
> *R.F.K. must die—R.F.K. must be killed Robert*
> *F. Kennedy must be assassinated R.F.K. must*
> *be assassinated R.F.K must be assassinated . . .*
> *I have never heard please pay to the order of of of*
> *of of of of of of of of this or that please pay to*
> *the order of*[6]

The notebooks clearly revealed the mind of a seriously disturbed person, but there were also what one reporter called "baffling incongruities" about Sirhan on the stand. "One minute he will seem ingenuous; the next, frightening," wrote *New York Times*'s Lacey Fosburgh. "And for this reason, one of the curious things about this historic trial is that there is a mystery. No one awaits a secret witness or the disclosure of the real murderer. But gradually developing more each week is a mystery about Sirhan himself: Who is he? What is he?" And why had he turned his rage on Robert Kennedy when other candidates—Eugene McCarthy, Richard Nixon, Hubert Humphrey, to name but three—had been far more outspoken in their support for Israel?

The same question was also troubling defense team investigator Bob Kaiser, who suspected that the answer might be linked to the references to money in Sirhan's notebooks. The more he thought about it, the more it seemed to him that money had changed hands, or been promised, and Sirhan had been hypnotized to forget it. "You wrote certain things over and over and over again," he pressed Sirhan. "And when you wrote about killing Kennedy, you join to it the unexplainable phrase, 'I have never heard Please pay to the order of of of of of.' "[7] But Sirhan stonewalled: He couldn't remember writing it; he couldn't even remember the notebooks, he insisted.*

"We know less than 1 percent of what really happened in history, but we know even less about what is happening around us right now,"

---

* At one point Sirhan said that he had focused his rage on Robert Kennedy because of the senator's support for the sale of fifty Phantom jets to Israel. But Kennedy had urged the U.S. government to send Israel the "fifty Phantom jets she has so long been promised," at a speech at Temple Isaiah in Los Angeles on *May 21,* and Sirhan had written of his intention to kill him on *May 18.* At his trial, Sirhan claimed that his rage was caused by a television documentary he had seen "about three weeks before the assassination." The problem with this, as with his earlier explanation, was that the documentary aired May 20—two days *after* the entry in his notebook declaring his determination to kill Kennedy. He then suggested a third explanation: "It was the hot news on this KFWB station where the announcer said that Robert Kennedy was at some Jewish club in Beverly Hills, or Zionist club, or whatever it was; and he had committed himself to formally giving or sending fifty jet bombers to Israel." (Trial transcript, pp. 4975–4976.) But the broadcast had aired even later than the television program.

Mohammed Ibrahim says Khaled Hassan complained to him in Damascus during the Sirhan trial.[8] An intelligence officer, Ibrahim had the trust of the most senior conservatives, as well as access to some of Arafat's most extreme hardliners in Fatah, and Ibrahim knew how disquieted the trial was making those on both sides of the divide who were aware of Mahmoud Hamshari's secret visits to Los Angeles. For although the individual elements might be inconclusive, even somewhat ambiguous, when taken together—timing, location, opportunity, rumor, not to mention Hamshari's call, in the desolate aftermath of the Six-Day War, for a top-level assassination on American soil—they were difficult to dismiss completely.

In order not to stir what was clearly a potential wellspring for even more internecine division, there was a rare consensus in Fatah to refrain from public comment on Sirhan, who had become the symbol of Arab terrorism in the West. Even so, Abu Iyad—whose intelligence operation remained a law unto itself, and out of the reach of any Fatah oversight—regarded the trial as an opportunity to push the Palestinian cause. However, distributing tens of thousands of posters featuring Sirhan's picture over his defiant war cry after he shot Kennedy—"I did it for my country"[9]—and claiming him as an Arab hero, was far too hubristic for many of Fatah's conservatives.

It had been widely rumored that Sirhan was linked to Fatah, and even though the suspicions faded somewhat when both legal teams in Los Angeles ruled out conspiracy, the missing months—the impenetrable "white fog"—in Sirhan's life coincided disturbingly with Hamshari's trips to California.

Abu Iyad, nevertheless, routinely denied that the assassination had been a Fatah operation, and although this does not necessarily mean that he was completely disassociated from a rogue operation mounted by Hamshari, Ibrahim feels there is some reason to believe him on this. "Abu Iyad was a professional; he would have made sure that Sirhan would not have survived to be put on trial," he says.

It is true that the moment he pulls the trigger an assassin has outlived his usefulness; and by the simplest order of reasoning, Sirhan should have been taken out by Kennedy's LAPD protection officers before he could get off a second shot. So had Abu Iyad been so sure that Sirhan would be gunned down on the spot that he had decided not to risk putting in a second Fatah gunman to take care of him in the predictable panic following his attack?

If this is correct, it was unfortunate that on that night, Kennedy had dismissed his LAPD protection team, and none of his own bodyguards was armed. "According to Kennedy's security man, Bobby didn't want any of our people around him because his constituents were basically black and liberals who didn't like policemen," former LAPD detective Bill Jordan recalled of Kennedy's concern for political correctness.[10] It had probably cost him his life, and saved his killer's.

In the meantime, it was the ferocity of Sirhan's obsession with killing Robert Kennedy, the proof written in his own hand, over and over again, that grabbed the headlines. But one sentence in his notebooks was skipped over during the trial. It read, "God help me . . . please help me. Salvo Di De Salvo Die S Salvo,"[11] an apparent reference to Albert DiSalvo, whose murder of thirteen women between 1962 and 1963 earned him the epithet "the Boston Strangler."

But despite his extraordinary *cri de coeur*, and even though it had come up and DiSalvo's crimes were discussed, in a recorded conversation in Sirhan's cell on the night he was arrested Sirhan claimed that the name DiSalvo meant nothing to him.[12] Convinced that the murders had been committed "just recently," Sirhan became agitated when told they had happened five or six years previously. "No, no, no, no," he insisted. "This was, like, last . . . nearly a year."[13]

Listening to the audio tape later, defense investigator Bob Kaiser felt that Sirhan had "obviously read enough about the Boston Strangler" to know that DiSalvo had gone about his business "in some kind of dissociated state."[14] Although the connection was not pursued, or not pursued carefully enough, it made Kaiser wonder why, if Sirhan had programmed himself to kill Kennedy—as his defense psychiatrist Bernard Diamond claimed—he had no apparent recollection of his autohypnotic sessions.

Was it possible, Kaiser wondered, that somebody else, perhaps one of the people with whom Sirhan had studied the occult—perhaps somebody he had met during his months in that "white fog" which the LAPD had been unable to penetrate—had programmed Sirhan to kill Kennedy, "possibly without his knowledge?" [15]

Earlier, when Diamond saw how quickly Sirhan could be hypnotized, and realized that he had been hypnotized frequently before, Diamond put him into a light trance, gave him a yellow legal pad, and told him to write something about Kennedy. Sirhan wrote, "RFK RFK RFK RFK RFK." Diamond asked him to write more than Kennedy's name, and Sirhan wrote, "Robert F. Kennedy. Robert F. Kennedy." What about Kennedy? Sirhan wrote, "RFK RFK RFK RFK RFK must die RFK must die RFK must die," nine times, until Diamond told him to stop.

The experiment, witnessed by Kaiser and Dr. Seymour Pollack, director of the University of Southern California's Institute of Psychiatry and Law, who represented the district attorney's office, had been continuing along these lines for a while when Diamond asked whether Sirhan thought he was "crazy."

"No no no no," wrote Sirhan, still in a trance. If he wasn't crazy, why was he writing in such a crazy fashion? "Practice practice practice practice practice," Sirhan wrote. Practice for what? "Mind control mind control mind control mind." Mind control for what? "Self-improvement self improvement," Sirhan wrote. Was he hypnotized when he wrote his notebooks? "Yes yes yes." Who had hypnotized him? "Mirror my mirror," Sirhan wrote. "my mirr my mirror." Who had taught him to do this? Sirhan wrote down AMORC AMORC. It was the acronym for the Ancient Mystical Order of the Rosae Crucis, the society he had joined after emerging from the "white fog." Did they show him how to do it? "No," Sirhan wrote. Did he learn how to hypnotize himself from a book then? "Yes yes yes." Who gave him the books? "I bought the books I bought."

They were going round in circles. But had Kaiser been on the right track when he wondered whether the Palestinian had been programmed—*"possibly without his knowledge"*—to kill Kennedy? The idea still sounded to him too much like *The Manchurian Candidate,* Richard Condon's Cold War thriller about a Sino-Soviet plot to turn an American soldier captured in Korea into a remote-controlled assassin programmed to kill the President of the United States. Although he was

skeptical, Diamond set up an experiment in which he would hypnoti-
cally program the Palestinian to climb the bars of his cell like a monkey.

Discovering himself one morning during the trial clinging to the top
bars of his cell, and suffering a mental block about his session with Diamond, Sirhan became "struck with the plausibility of the idea" that he
had been programmed to kill Kennedy.[16] Although the experiment shed
no more light on who might have been his programmer, Kaiser continued to suspect that "somewhere in Sirhan's past there was a shadowy
someone."*

He later complained, "With a little more diligence than they exercised, and a great deal more intelligence than they had, the police might
have established links between Sirhan and the right wing, between
Sirhan and the left wing, *between Sirhan and the Al Fatah . . .*"[17]

There seem to be two reasons why the Manchurian Candidate theory
was not pursued more avidly at the time. The first is that a political
motive had been ruled out by both the defense and prosecution. The second reason, clearly, is that it was easier to believe that Sirhan was a loner
with a grudge and a gun than it was to believe in hypnoprogrammed
political assassins.

Even so, it is unfortunate that Sirhan's references in his notebooks to
Albert DiSalvo were not pursued more strenuously; if they had been,
perhaps the reason for Sirhan's unusual interest in a case in which hypnosis had played a vital role in convicting the killer might have been discovered and may have made those who dismissed Kaiser's theory think
again.

---

* Kaiser turned up several interesting real-life examples of murder-by-proxy, through
hypnosis, including a case in Copenhagen in 1952 in which Bjorn Nielsen had programmed Palle Hardrup to rob banks and kill anyone who got in his way in order to
raise money for a new political party seeking the unification of Scandinavia. Nielsen
was convicted and sentenced to life imprisonment for having "planned and instigated
by influence of various kinds, including suggestions of a hypnotic nature," the commission of robbery and murder by another man. (Paul J. Reiter, *Antisocial or Criminal Acts
and Hypnosis.* Springfield, Ill: Charles C Thomas Co., 1958.)

*   *   *

Boston lawyer F. Lee Bailey had successfully prosecuted the Albert DiSalvo Boston Strangler case with the assistance of a remarkable clinical hypnotist and Los Angeles physician named Dr. William Joseph Bryan, Jr.

A hypnosis superstar, few doubted Bryan's claim that he had worked on the CIA's mind control program, MKULTRA.[18] A man who could make subjects bleed on cue[19]—as he had done at a hypnosis seminar for trial lawyers in San Francisco in 1961—was unlikely to have been overlooked in what had been called the CIA's "brain war."[20]

Nevertheless, because it was a project that the Agency itself admitted could not be run under normal ethical conditions—and accepted that "the best scientists in this field would be most reluctant to enter into signed agreements of any sort"[21]—no contracts ever existed, and Bryan's role in MKULTRA is impossible to verify.

But a woman who knew Bryan well at that time (she had been his mistress, and wishes to remain anonymous), recalled that he had told her he worked for the CIA, for a man named Gottlieb, who eventually fired him—because, Bryan claimed, he had moonlighted as technical adviser on the *Manchurian Candidate* movie.

Gottlieb was probably Dr. Sidney Gottlieb, who ran the MKULTRA program and, in 1954, after a year in development by a CIA behavioral research team code-named ARTICHOKE,* took over the "ultimate experiment in hypnosis"[22]—the creation of a programed assassin: the Manchurian Candidate. Exempt from most internal CIA controls, its most secret records destroyed, lost, or doctored,† it is not known

---

* On February 19, 1954, Morse Allen, a CIA behavioral researcher, having put a secretary into a deep trance, hypnotized a second secretary and told her that if she could not wake up her friend, "her rage would be so great that she would not hesitate to 'kill.' " Even though she had earlier expressed a fear of firearms of any kind, she picked up a gun planted by Allen and "shot" her sleeping friend. After the "killer" was brought out of her trance, she had apparent amnesia for the events, denying she would ever shoot anyone. (John Marks, p. 183, citing CIA memos acquired under the FOIA.)

† A surviving CIA audit of the project in 1963 reemphasized the need for secrecy: "Precautions must be taken not only to protect operations from exposure to enemy forces

whether a successful Manchurian Candidate was achieved, or tested operationally, by the time the project was terminated in 1964—twelve months after Bryan was fired from the program by Gottlieb.*

"Being around Bryan was an extreme experience," said his former assistant Linda Gordon. An obsessive, brilliant, overweight, impressive-looking man with the etiolated skin of a person who rarely ventures outdoors, "he was a mess as a human being, but he would have dared anything if his ego was challenged," Gordon told me.

Bryan, of course, was the man David Karr had sent Hamshari to when the Palestinian complained that the smog in Los Angeles gave him migraines; and the man whom Onassis visited in Las Vegas, to cure his insomnia—or perhaps his unreliable potency—on the eve of his pre-honeymoon cruise with Jackie in the spring of 1968. It was a tangled web but through it like a thread ran the names of Onassis, Karr, and Bryan—and now Sirhan Sirhan, whose fascination in extreme hypnotism, and Bryan's genius, could finally be understood.

On April 17, 1969, after sixty-four sequestered days and nights, sixteen hours and forty-two minutes of deliberation, the jury of seven men and five women found Sirhan Bishara Sirhan, "alone and not in concert with anyone else," guilty of murder in the first degree. He was sentenced to death in the gas chamber.†

---

but also to conceal these activities from the American public in general. The knowledge that the Agency is engaging in unethical and illicit activities would have serious repercussions." (Quoted in Church Committee report, FR1, p. 394.)

* "It cannot be done by everyone, it cannot be done consistently, but it can be done," says Milton Kline, former president of the American Society for Clinical and Experimental Hypnosis, and one of the outside experts with whom Gottlieb's scientists discussed the Manchurian Candidate project. (Marks, p. 187.)

† His sentence was later reduced to life imprisonment when the United States Supreme Court declared the death penalty unconstitutional.

TWENTY-FIVE

A SUICIDE WAITING TO HAPPEN

We know the truth, not only by the reason, but also by the heart.

—BLAISE PASCAL, 1623–1662

I n 1970, a tragic subplot unfolded on Stavros Niarchos's private island of Spetsapoula.

Shortly after seven o'clock on the morning of Monday, May 4, 1970, Onassis got a call from his security chief Miltiadis Yiannakopoulos telling him that Niarchos's wife, Eugenie, was dead. She had been found by her maid lying unconscious on her bedroom floor with an empty bottle of Seconal by her side.

In a statement issued to the press later that day, Niarchos explained that his wife had been distressed by a conversation she had overheard him having on the telephone with his ex-wife Charlotte Ford in New York, in which he invited their four-year-old daughter, Elena, to spend the summer with him on Spetsopoula.*

---

* Elena had been conceived in a strange interlude in 1965 when Niarchos married automobile magnate Henry Ford's daughter in a Juarez motel suite twenty-four hours after

The tensions had been close to the surface of the Niarchos marriage since Stavros returned to the family fold after his short-lived marriage to the Ford heiress. Eugenie had never been a robust woman, and she had been deeply depressed at Niarchos's refusal to remarry her (he insisted that in the eyes of the Greek Orthodox church his marriage to Charlotte had never happened). For months friends had been expressing concern about her emotional fragility and her barbiturate habit; she was "a suicide waiting to happen," said one.

Nevertheless, Niarchos's behavior between 10:25 P.M. on May 3, when Eugenie was found unconscious, and 12:25 A.M. the following morning, when she was pronounced dead, would arouse suspicions that exist to this day. It was more than half an hour before he summoned a doctor—not, as you might expect, the doctor from the island of Spetsapoula, a ten-minute motorboat ride away, but the Niarchos company physician in Athens, who arrived by helicopter about ninety minutes later.

Nevertheless, her death was not from natural causes, and there were bruises on Eugenie's body. Because of this, the company doctor refused to issue a death certificate, and the body was flown to Athens for a post-mortem examination. Niarchos would later try to explain away the bruises to the Piraeus public prosecutor by saying that he had shaken and slapped his wife in an attempt to revive her; that she had fallen several times as he struggled to get her back on the bed, and that he had taken her by the neck to haul her upright.

The autopsy report cited a two-inch bruise on the abdomen with internal bleeding, and bleeding behind the diaphragm in the area of the fourth and fifth vertebrae; a bruise on the left eye and swelling on the left temple; an elliptic hemorrhage on the right side of her neck; a hemorrhage to the left of her larynx with contusions above the collarbone on

---

Eugenie had divorced him in the same Mexican town on the grounds of incompatibility. But even before Charlotte returned to Juarez, less than two years later, to divorce him as summarily as she had married him, Niarchos had found his way back to Eugenie and their four children.

the left side of her neck, her left arm, ankle, and shin; and some news reports claimed that she had also suffered a ruptured spleen. Nevertheless—after a slap on the wrist for Niarchos's "inexplicable and inexcusable delay in calling a doctor"—it was concluded that Eugenie's injuries were consistent with strenuous attempts at resuscitation, and that the death was caused by an overdose of barbiturates.

Not everybody was satisfied with this verdict, however, and a second autopsy was demanded by an investigating magistrate. This time, it was ruled that the Seconal found in her body (two milligrams of barbiturate in one hundred cubic centimeters of blood) was not a lethal dose—and that Eugenie had died as a result of her injuries.

"Stavros was in more trouble than he had ever been in in his life," said Niarchos's former public relations man Alan Campbell-Johnson, who had done so much to help Niarchos destroy Onassis's Jiddah deal fifteen years before. "It would be disloyal of me to say that Stavros had only got what he deserved, but many people believed that to be so."[1]

And there was far worse to come.

Cannily, Onassis had transcripts of all the telephone calls made to and from Spetsopoula on May 3, and he knew that Niarchos's account of events on the evening Eugenie died was a complete lie. Not long after she was buried at the Bois-de-Vaux cemetery in Lausanne, he summoned Georgakis to avenue Foch and, in the presence of Johnny Meyer, played the tape of the conversation between Niarchos and Charlotte Ford, which Niarchos claimed as the cause of his wife's suicide.

It was plain to Georgakis that Niarchos had been drinking. His voice, which was never exactly mellifluous—as he affected an upper-class English accent, it often sounded so strangulated that people sometimes believed he had a speech defect—was slurred and his conversation repetitious.

Onassis wanted to know whether Georgakis could hear Niarchos's daughter Elena being discussed. Georgakis listened to the tape twice; Elena's name wasn't even mentioned. (According to the writer Sally Bedell Smith, at the end of the conversation, Charlotte Ford *"hung up, nonplussed."*[2])

Onassis then played a second tape, which he claimed had been recorded on the same day, of a conversation between Niarchos and his sister-in-law Tina. Still unmistakeably drunk, and indiscreetly amorous, Niarchos was pleading with Tina—now married to the marquess of Blandford, heir to the Duke of Marlborough—to spend the summer with him on Spetsopoula.

This was the call Eugenie overheard. *This* was the reason why Eugenie wanted to kill herself, Onassis said triumphantly in the stunned silence that followed the end of the tape: *"She found out that Stavros was fucking her little sister!"* [3]

Meyer was not surprised that Onassis had a tape of the conversation. "Not a sparrow fell on Spetsapoula without Onassis knowing about it," he later told Brian Wells.[4] But Georgakis, who had always believed that the bitterness that existed between Onassis and Niarchos could only be settled "not by justice, not by law, but by power and revenge, by methods entirely in keeping with their characters," knew what Onassis had in mind, and was appalled.[5]

A few days earlier, following the second damning autopsy, the investigating magistrate in Athens had recommended that Niarchos be charged with "inflicting injuries leading to the death" of Eugenie, and be tried under Article 311 of the Greek penal code: involuntary homicide, which carried a maximum penalty of eighteen years in prison. The indictment had gone to the High Court for approval.

"I said, 'Ari, if this tape gets out, it'll do for Stavros,' " Georgakis recalled telling Onassis. " 'But it will hurt the kids far more. They've lost their mother, do you want them to lose their father, too?' "[6]

There are conflicting opinions about what happened next. According to Meyer, Onassis told him to leak the tapes to the *New York Post* anyway,[7] and changed his mind only when Onassis's security man Miltiadis Yiannakopoulos warned him that they would lose a precious intelligence source if Niarchos discovered that his telephones on Spetsapoula were tapped.

But according to a source in the Niarchos family, Onassis backed off when Niarchos threatened to retaliate with "some serious shit" he had on him. "They both had skeletons in their cupboards; getting the goods on each other was what had always kept the balance of power. But I always felt that Stavros had rather more on Ari than Ari ever had on him. And when you consider what Ari had on Stavros, you have to wonder what Stavros could have done that topped that," said my source.*

Meanwhile, a third postmortem on Eugenie was arranged. This time, the report—written by four doctors whose verdicts were already in conflict, plus two professors of morbid anatomy, and two pathologists nominated by Niarchos, and owing more to the skills of diplomacy than the exigencies of forensic medicine—got Niarchos off the hook.

But although there were still too many unanswered questions for the rumors of Niarchos's culpability to go away, Onassis suddenly became strangely silent on the matter, confirming the suspicion of those who believed that Niarchos must have "some serious shit" on him.

In the spring of 1971, Onassis decided that Christina should marry Peter Goulandris, whose family owned four shipping lines, and whose mother was a member of the distinguished Lemos family. But though Christina liked Goulandris well enough, the day before their engagement was to be announced, she fled to Monte Carlo where, by the swimming pool of the Hotel de Paris, she met Joseph Bolker.

Although the relationship quickly deepened on Christina's part to a degree of dependency, it was essentially a sexual one for the forty-eight-year-old American, and he was far from happy when she followed him to Los Angeles. His apartment on the twenty-fourth floor of Century Towers West was a perfect bachelor setup, and he did not want to cramp his style with an uninvited live-in mistress, even if her father was one of the

* Niarchos's eldest son, Philippe, ruled that none of the family nor their associates should talk to me or help me in any way after he had found a letter I had written to Stavros shortly before his death outlining the premise of this book and asking for an interview. My source, who had talked to me before the edict was issued, pleaded with me not to mention his name or to use anything he told me that could be traced back to him.

richest men on earth, and married to the widow of the thirty-fifth president of the United States.

Tina inflamed the situation by declaring that she did not want her daughter living in sin, and Christina must return to Europe at once—"or make it legal." Seeing her lover's reluctance to marry her as a slur on her desirability, Christina took an overdose. Saved by a young doctor who lived across the hall, she was unrepentant. "I will just keep doing this until you marry me," she told Bolker.[8]

On July 28, Jackie was celebrating her forty-second birthday on Skorpios when Onassis heard that his daughter had married Bolker in Las Vegas the previous evening.

One of the unfortunate paradoxes of Christina's character was that although she wanted her independence, she was also addicted to a lifestyle that only daddy's money could buy. She was distraught therefore when he promptly rewrote the terms of her trust fund—she had been due to collect $75 million on her twenty-first birthday in December—postponing indefinitely her access to the fortune that would set her free.

Dependent on her husband—a man whom she had described as a "dinky millionaire" in property—she would miss Daddy's private island, his jet planes, the charge accounts, and permanent suites in the best hotels around the world. If the price of life with these luxuries meant being beholden to her father, the price of life without them would be even more insufferable.

But in the middle of divorcing the marquess, Tina summoned her daughter to New York. To help make amends for the blocked trust fund, she gave Christina $200,000 (an extraordinarily generous gift from a woman notorious for her thrift: "To get Tina to break into a five-pound note was a goddam miracle," said Fiona Thyssen[9]), and urged her to stand up to her father.

The money, however, like the advice, was given not entirely without self-interest; and when, eighteen months after Eugenie's death on Spetsopoula, Tina and Niarchos were married in Paris, Christina suspected that her mother had financed her rebellion simply to divert Onassis's attention from her own plans.

Bolker said he had never witnessed such a goddam Greek tragedy. "How could her mother marry this man who had beaten her sister to death and forced the barbiturates down her throat to make it look like suicide? That's what Christina believed. She said, 'You must understand, Joe, men like Niarchos, and men like my father, consider themselves above the law. What other people think or want doesn't count,' " Bolker told me.[10]

Christina's accusations became common gossip in Athens and Paris; another ritual overdose testified to the sincerity of her anguish. Tina told Bolker that he must stop Christina "telling these monstrous lies about my husband."[11]

These people were unlike anyone Bolker had ever met. "These families carved up countries between them: They owned judges, prosecutors, politicians, cops," he said, and refused to get involved. "I said, Tina, If you can't stop Christina saying these things, how do you think I can?' " he later recalled his dilemma. Christina was convinced that Stavros had killed Eugenie, and there was nothing Bolker could do about it. If it was just a suspicion, he told his mother-in-law, they could discuss it rationally. But it wasn't a suspicion: It was an obsession—and "nobody can discuss an obsession rationally."[12]

Tina told him that Onassis had also done "dreadful things in his life. Wicked, wicked things." Christina must simply learn to be discreet, or suffer the consequences, she said.[13]

Bolker had stopped being surprised by anything Tina and her daughter said. Onassis was "capable of having people killed," Christina had once told him—or at least, she backed off a little when she saw the fear in his eyes, "having their legs broken or something."[14]

When she arrived at Fiona Thyssen's house on Wilton Place following a lunch with her mother at Claridges in December 1971, Christina's face was wan and she sat down so shakily that her brother's mistress thought she was going to collapse. "My mother seems to enjoy finding new ways to hurt me," she told Thyssen miserably.

Although she was by nature garrulous and confiding—"Christina was a blurter," recalls one of her English lovers—she knew she could trust nobody with the story her mother had told her that afternoon to make her stop telling those "monstrous lies" about Stavros.

She dared not even confide in Thyssen. Young enough still to be fun but old enough to be able to talk from experience, her brother's mistress was the wisest friend she had. She admired the way Thyssen had handled her own difficult times, and the way she had encouraged Alexander to stand up to their father's tyranny, and had brought about Christina's reconciliation with her brother after years of mutual jealousy and suspicion. Even so, the story Christina had been told that afternoon had disturbed her too much even to confide in—and warn—the woman whom she loved as a sister, and whose life had become so desperately mixed up in theirs.

Shortly after Alexander arrived in London the following day, he had a private conversation with his still distraught sister at Wilton Place.

Although Tina's story might have been dismissed as further proof of her madness—insanity being the only explanation her children could find to account for her marriage to her sister's killer, as they believed Niarchos unquestionably was—neither of them underestimated their father's ruthlessness, or could ignore the apparent irrefutable evidence their mother had given Christina at their lunch at Claridges.

It was Alexander who now took charge of the situation. Apart from the startling evidence Tina had produced to support her story, we can only speculate about what he put in the file he eventually placed in a safe deposit box at the Plaza-Athenée hotel in Paris.

"Alexander told me that the file contained information that would dissuade his father from allowing harm to come to me," recalled Thyssen, who did not press him for details. "I knew he would tell me only what it was safe for me to know," she said.[15]

Shortly before this episode, Thyssen had inadvertently seen a transcript of a conversation between two of Onassis's "hatchet men" that made it clear Onassis was "determined to do a number on Joe [Bolker], he wanted to hurt the fellow, perhaps not do him in, but certainly do him some serious harm."[16]

She had always believed that there was a vein of madness in Onassis, which made him a dangerous man to cross. "He would stop at nothing to get his way . . . he was capable of anything if he felt threatened. I would have been extremely stupid if I had not realized that the things I knew about him put me in considerable danger," she told me. "Alexander and I were aware that accidents could happen. . . ."[17]

Meanwhile, Christina continued to worry about the deteriorating state of her marriage. "She was sensational-looking in those days, very bad table manners, like most Greeks, but a sensational face," said Thyssen, who found it painful to see someone so young so afraid of the future.[18]

Aware that unless Christina acted fast, Bolker would get a serious beating, and maybe something worse, Thyssen urged her to end the marriage as quickly as she could. "I said, 'Look, you don't want to be married to Joe. He doesn't want to be married to you. Go back to California, tell him you want a divorce, and in a few weeks it will be all over and nobody need be involved except you and Joe. Do something with dignity, which is a word your father doesn't even begin to understand.[19]

On December 9, Christina returned to California for her twenty-first birthday party two days later at the Bistro in Beverly Hills. The following day, she and Bolker agreed to divorce. "You can have no idea how relieved I felt to be finally getting out of that family," Bolker told me.

A couple of months later, Christina was staying at the Copacabana Palace in Rio de Janeiro when her father's Learjet, flying to meet him in Nice, crashed into the sea off Cap d'Antibes, killing his pilots, Dimitris and Giorgio Kouris.

Christina called Bolker and told him: "They're trying to kill my father, Joe." When Bolker asked who "they" were, she said that her father had once had somebody killed—"somebody big, or somebody important," Bolker recalled—and that his plane had been sabotaged in revenge. Previously, she had claimed that her father was "*capable* of having people killed."[20] This was the first time she had said that Onassis had actually had somebody killed. Alarmed at the things she was saying on an open line—and aware that there was usually a vague element of truth in all her stories—Bolker ended the conversation quickly.*

* I first met Joseph Bolker in February 1983, while researching *Ari*. In this first interview at his home in Los Angeles, Bolker mentioned Christina's remark about her father

Although the crash—on a night approach to Nice airport, a few
weeks after David Karr's yacht, *Asmeda Hope,* had mysteriously exploded
and sank in the harbor at Cannes—was blamed on pilot error, Onassis
was convinced that, like Karr's yacht, the plane had been blown up. The
pilots' brother Grigoris Kouris also believed that his brothers had been
murdered in an attempt on Alexander's life. Had they not been delayed
for nearly an hour leaving Athens to pick up Alexander in Nice to fly
him to Paris, "Alexandros would have been on board when the bomb
exploded," he said.[21]

Alexander, who had been close to the Kouris brothers—they had
taught him to fly and were like brothers to him—spent weeks with Fiona
Thyssen searching the French coastline for traces of the wreckage. But
when it was finally located on the seabed off Cap d'Antibes, Onassis
refused to have it raised.

"Why didn't Ari want that plane salvaged?" Thyssen later pondered.
"Why didn't he want the wreckage forensically examined? He was convinced the plane had been sabotaged, yet he wouldn't pursue the investigation. What was he afraid they would find?"[22]

Even before Tina's accusations, Alexander knew that his father was a seriously unbalanced man. Because there were no oversight committees to
curb his excess, and no one around him with sufficient power or courage
to second-guess him, his vast wealth allowed him to make decisions that
often exceeded normal ethical limits. Understandably, he was a man who
had made plenty of enemies. Alexander had grown up knowing this, as
surely as he knew that he would never be happy until he broke free of his

---

once having had somebody killed—"somebody big, or somebody important"—but
later asked me not to print it. Eventually, he agreed that I could write that Onassis was
"capable of having people assassinated . . ." Shortly before he died of cancer, aged 62, in
1986, I asked him whether he knew the name of the man Onassis had allegedly had
murdered. He said, "Christina never gave me a name and frankly I never wanted to
know. What I do know, and have no doubt about, is that she was telling the truth.
Onassis did have somebody important killed. And in my opinion, all Christina's problems and all the tragedy that's befallen that family started with that killing."

tyranny. "It's the only way I'm going to survive, I can't take this grotesque man's domination for much longer," he had told Fiona Thyssen.[23]

And now he faced the possibility that two of his dearest friends had died in a bungled attempt to kill him in order to punish his father. This troubled Alexander's conscience more than it had ever been troubled before. It was not surprising, then, that, a few weeks after the crash, he confided his mother's extraordinary story to Yannis Georgakis and revealed what had passed between Tina and Christina at Claridges:

His father had financed the assassination of Bobby Kennedy—and might also have paid to have Fiona Thyssen and Stavros Niarchos murdered as well!

Before Georgakis could protest, or perhaps laugh, Alexander handed him an envelope containing photostats of pages from Sirhan's notebooks, which Tina had given to Christina at Claridges Hotel in December.

Even three years on, anyone who had followed Sirhan's trial in the newspapers would probably have recognized the maniacal ranting and compelling streams of consciousness, jumbled up with political mantras, race track intelligence, and reminders of his mission to murder Bobby Kennedy, found in the notebooks that had helped convict him.

On the first page, Sirhan had written at the center of a roundel, amid Arabic writing, the single name,

Fiona.

And on another page:

2 Niarkos!

On a third page, between the lines *One Hundred thousand Dollars and Dollars—One Hundreds,* Sirhan had written in Arabic: *they should be killed.* And next to that, the number:

Three.

Fiona, Niarchos, and Kennedy: The names were startling by virtue of their very juxtaposition. But, as a lawyer, Georgakis was always skeptical: he did not trust facts that were startling, and circumstantial evidence made him uncomfortable. But three names and a sum of money written in a killer's notebook—he had seen far flimsier evidence than that get a conviction in a court of law. The truth was, he was *shattered* by the very real possibility that Onassis had paid for the bullets that killed Bobby Kennedy, and not for the first time he wondered about the nature of Onassis's deal with Hamshari.

"Sometimes remarkable chains of circumstances can reasonably be ascribed to coincidence. But at some point the connection is so astonishing that one must assume that it could only have happened by design," Georgakis would later muse. "And when an Irish-American name, a Scottish name, and a Greek name—the names of Onassis's three most loathed and troublesome enemies—are found in a Palestinian killer's notebook written in Los Angeles shortly after Onassis had given more than a million dollars to a Palestinian terrorist in Paris [far more than that Palestinian had demanded], and the first name on that list has already been killed, you must seriously consider the possibility that you have reached that point." [24]

TWENTY-SIX

ANOTHER SAD SUGAR DADDY

**Man imagines that it is death he fears;
but what he fears is the unforeseen . . .
What man fears is himself, not death.**

—ANTOINE DE SAINT-EXUPÉRY, 1900–1944

Alexander could never forgive his father for what he had done, although he pitied him, too. Onassis's marriage—which had been at least partly conceived as a business gambit, and for which he had risked so much, and paid so dearly—had brought him nothing but misery. Jackie had not saved Omega for him; she was high maintenance, and instead of giving him a touch of class and opening doors for him in Washington, the marriage had anathematized him across America. In fact, Alexander told a Paris friend, Jackie made his father look like "another sad sugar daddy."[1]

Onassis knew it, too. "My God, what a fool I have made of myself,"[2] he told friends when asked about the woman he had once lyrically compared to a diamond—"cool and sharp at the edges, fiery and hot beneath the surface"[3]—and now dismissed as "coldhearted and shallow."[4] He no longer boasted of her imagination and ardor in bed, or enjoyed repeating the obscenities he claimed she articulated during apogees of lovemaking. Nevertheless, he continued to summon her from New York, as if she were one of Madame Claude's girls, just for sex, and dismissed her when he was satisfied or bored by her. Christina told friends that it was his way of reminding her "what she was."

He had slashed her $30,000 monthly allowance by a third and moved its control from New York to his Monte Carlo headquarters, where he could keep a tighter grip on her spending. But whether her extravagance was the revenge of a wife who knows that her days are numbered or something else is hard to judge. "Some kind of battle was going on in that marriage that was wonderful to watch and fascinating as hell to think about," says the painter Rico Zermeno, who had known Jackie since 1949, when she began her Sorbonne year in Paris. Zermeno knew that, in spite of Onassis's humiliation of her, she would not just fold her hand.[5]

But although it did not bother her that her husband and Callas had resumed their affair, she did not want him romancing his mistress in public, she told Billy Keating. After all, she was discreet—she had revived her affair with Roswell Gilpatric, but she was usually accompanied in public by an acknowledged gay—and she expected her husband to show her the same consideration.

But Onassis had too much on his mind to care about Jackie's finer feelings. After Moscow had offered him "virtually unlimited credit, equipment, and know-how"[6] to develop Omega, the colonels—told by Washington that if they let the Soviets into Greece through the back door, they could forget any more U.S. aid for their regime—cancelled the whole project.

Alexander, meanwhile, was turning the tiny Olympic Aviation offshoot of Olympic Airways into a nicely profitable operation. Unlike his cousin and stepbrother, Constantine Niarchos, whose bad-boy excesses were the despair of his family, Alexander was a son who would have made any normal father proud. But Onassis was not a normal father, and Alexander knew that the contents of his safe deposit box at the Plaza-Athenée hotel in Paris had damaged their relationship, probably forever. Onassis would

spend hours eviscerating him on the telephone; his scorn and threats
became a daily torment. Unable to convey to Fiona Thyssen the extent of
his rages, Alexander began taping their phone conversations. "Listen to
him," he told her, playing one tape. "It's two o'clock in the afternoon over
there [New York], and he's completely pissed out of his mind."[7]

It was clear to them that they could not make a future together that
was so vulnerable to Onassis's tyranny. Alexander must break free.
Thyssen bought a house in Switzerland, and Alexander began to make
plans to make a new start beyond his father's reach. "The plan was that
as soon as the house was ready he would say to his father, 'Right, I'm
going to live with Fiona; I'm going to university; I'm going to get a
degree, and I'm going to get a job.' He actually had the guts to do that,
to walk away and be totally disinherited," Thyssen said.[8]

In the autumn of 1972, Onassis stepped up his plans to dump Jackie. He
knew it would not be cheap, he told his New York lawyer Roy Cohn, but
he wanted to ensure that he didn't pay a penny more than he had to.

"Mr. Onassis said, 'I owe that woman nothing,' " Cohn recalled. "He
had definitely concluded that he wanted to break the marriage and had
been consulting his Greek lawyers and so on, and there were a lot of
complications over there. But largely he was concerned that Jackie knew
certain matters concerning his private and business interests that he
would not want to become public knowledge."[9]

It was a smart move hiring Cohn, who not only had a reputation as
"the toughest, meanest, vilest, and one of the most brilliant lawyers in
America,"[10] but who also hated the Kennedys, particularly Bobby, who
had never forgiven Cohn for beating him to the post of chief counsel on
Joe McCarthy's Senate subcommittee back in 1953. Cohn told me that
Onassis "felt he was being taken for a sucker." Jackie was "again sleeping
with Ros Gilpatric, and others," and it was agreed that Johnny Meyer
would arrange around-the-clock surveillance to get evidence that could
be used against her.[11]

Mahmoud Hamshari had by this time abandoned his shadowy past and
various aliases to surface as the PLO's "ambassador" in Paris. But a leop-

ard never changes its spots,* and Israeli counterintelligence officers continued to distrust him, believing that he was a key figure in the Fatah
splinter group Black September, which was responsible for the Munich
massacre in which eleven members of the Israeli Olympic team were
killed.

But as the Mossad investigation into his involvement in Munich
entered its final phase, a more insidious threat to Hamshari was growing
inside Fatah itself. According to Mohammed Ibrahim, what was rumored
to be an "incontrovertible source" in Paris claimed that Hamshari had mis-
appropriated donations made to Fatah by, among others, Aristotle Onassis.

Such accusations were not uncommon in the incurably fractious
Fatah, where bookkeeping was lax and often idiosyncratic, and where
there were always internecine scores to be settled. But since charges were
more likely to stem from personalities and rivalries than evidence, they
were usually investigated and sorted by middle-rank intelligence officers
as a matter of routine. On this occasion, however—maybe because the
source was considered to be so reliable; or because of the large amount of
money alleged to be involved; or perhaps because he already knew the
truth, and needed time to deal with the cat that had been let out of
the bag—the investigation was handled personally by Abu Iyad.

Although there is no means of knowing what was going through the
mind of Fatah's intelligence chief as he embarked on the investigation of
Hamshari, a political balancing act would clearly be required by him if
the Paris source proved to be correct. For although Hamshari's value to
Abu Iyad was felt rather than understood inside Fatah, few doubted that
the first PLO "ambassador" to France was—as he had always been—Abu
Iyad's man. In this respect, his innocence or guilt could become the ful-
crum from which the balance of power between the hawks and the doves
in Fatah might shift.

Meanwhile, on the morning of December 7, several weeks after Abu
Iyad began his investigation, and shortly after "ambassador" Hamshari

---

* Hamshari was believed to have been implicated in an assassination attempt against
Ben-Gurion in Copenhagen in 1969; and was also responsible for the explosion aboard
a Swissair plane that killed forty-seven people in 1970.

left for his office at the Arab League building on Boulevard Haussmann, two Israeli agents posing as telephone engineers entered his apartment on rue d'Alésia in the fourteenth *arrondissement,* where they fixed a sophisticated explosive device beneath his desk. The following morning, Hamshari took a call from a man asking to speak to the ambassador. "Yes, this is he," he acknowledged his courtesy title, of which he was inordinately proud. The next sound he heard was the whine of a hypersonic signal setting off the bomb.[12] Although terribly injured, he miraculously survived the blast, and was placed under police guard in intensive care in the Hospital Cochin on rue du Faubourg Saint-Jacques.

There is, and there will probably always remain, doubt about how much Abu Iyad knew of Hamshari's deal with Onassis. For like so many agents, Hamshari had become ungovernable, a man alone, whose habit of secrecy and lies had separated him not only from his enemies but finally also from his family and closest comrades, too. Nevertheless, even if he had been following orders when he put the squeeze on Onassis in Paris, it is inconceivable that this was part of a Fatah strategy to set in motion the events that led to Bobby Kennedy's assassination. For that could only have been the result of fate, and an extraordinary twist in Onassis's personal life which no one could possibly have anticipated when Hamshari first advocated the murder of a prominent American in the aftermath of the Six-Day War the previous summer.

Although these questions may seem peripheral to whether Hamshari had embezzled Fatah funds, Abu Iyad knew that they were not peripheral at all. But whether he simply seized an opportunity to get rid of a man who knew too much—for his own comfort as well as Fatah's—we shall never know. What we do know is a secret court martial was held in Hamshari's unavoidable absence in a suite on the top floor of the St. George Hotel in Beirut. Presented with the conclusions of Abu Iyad's investigation, it found Hamshari guilty of embezzling PLO funds, by diverting donations, principally from Onassis, to numbered accounts in Switzerland and Luxembourg.

Although loyalty can never be taken for granted in the bitter rivalries of revolutionary politics, Abu Iyad felt deeply betrayed. According to

Mohammed Ibrahim, a friend who spoke to him shortly after the court martial found him more distressed at Hamshari's personal disloyalty to him than his act of embezzlement. Nevertheless, even as his protegé hovered between life and death in Paris, his was the first signature on the execution order.*

But although he had built one of the best intelligence networks in the Arab world, the speed with which Abu Iyad uncovered the link between Onassis's cash and Hamshari's numbered accounts suggests something more than smart detective work. Could therefore the "incontrovertible source" in Paris who had greased the skids for Hamshari have been David Karr? And, if so, was he simply carrying out Onassis's wishes now that Hamshari had become a liability? Unfortunately, given Karr's reputation for extreme intrigue, and his expertise in manipulation, even today the answer to that question is elusive.

Relations between Karr and Hamshari had deteriorated in the four years since Karr had contrived to introduce the Palestinian to Onassis in 1968. Privately, Karr suspected that Hamshari had been responsible for sinking his yacht, and perhaps for blowing up Onassis's plane, too. "These guys have memories like flypaper for an insult," he had told Georgakis when they met for dinner in Paris in the summer of 1972.[13]

The nature of Karr's relationship with Hamshari had always been murky, but what prompted that remark—and why there was now bad blood between them—is, like so much else in Karr's life, puzzling.

---

* The other signatories were Muhammad Yusif Najjar (the Black September operations chief, who would himself be killed four months later by the Israelis), and Arafat's deputy and the boss of the PLO's military wing. Abu Jihad. According to the PLO's constitution, no Palestinian can be executed without a written order signed by the Chairman. The man with the executive responsibility for preparing execution orders and submitting them to the Chairman for approval was Abu Iyad. Whether on this occasion this formula was strictly followed is unclear. When historian Alan Hart asked how many such orders he had presented to Arafat for signature since 1967, Abu Iyad said, "To be exact I cannot tell you because I truly do not remember. But let me say for the record that the answer to your question is many." (Alan Hart, *Arafat*, p. 20. London: Sidgwick & Jackson, 1994.)

Nevertheless, by this time Karr had embarked on another of his many careers—as an arms dealer, whose specialty, according to the Israelis, was selling Soviet weapons to the PLO.[14] It was a dangerous game, especially the way Karr chose to play it. "David's Jewish, but that doesn't stop him selling [Soviet] weapons to the PLO. And he takes commissions from both sides . . . I wouldn't want to guess where that dishonorable bastard's going to end up!" Karr's one-time partner, the formidable American business tycoon Armand Hammer had told friends.[15]

It is likely that the sinking of Karr's yacht was at the root of his falling out with Hamshari, and the reason why he might have decided to blow the whistle on the Palestinian's undeclared income from Onassis. After all, he had the knowledge, he had motive and the connections, and it is difficult to imagine who else—besides Onassis—would have stood to benefit quite so much from Hamshari's removal. Except, of course, Fatah itself.

"The truth about Bobby Kennedy's death was still a mystery, probably even in Abu Iyad's mind," says Mohammed Ibrahim, who believes that the Fatah intelligence chief was secretly relieved to finally be given an excuse to remove the source of that lingering doubt.[16]

On January 9, 1973, thirty-two days after he was blown up—and just when he appeared to be making a remarkable recovery from his terrible injuries—Hamshari developed a mysterious fever, and within twenty-four hours he was dead. Abu Iyad sent a medico-forensic team and five intelligence officers to Paris to investigate the death. "I am convinced it was a medical assassination," a PLO spokesman later told reporters, putting the blame at the Israelis' door.[17]

"Sure it was a medical assassination, but the Palestinians played Doctor Death, not us," said a Mossad agent, who admired the way Fatah even "got mileage out of Hamshari's corpse"[18] by treating him like a war hero.

On January 12, 1973, Alexander dined with Onassis at Fouquet's, on the Champs Elysées. Afterwards, Alexander called Fiona Thyssen and told her that "the old man" had been in terrific form; it was as if "a great weight had been lifted from his shoulders."[19]

Onassis had decided to dump Jackie, and to mark the occasion, he promised to buy his son a helicopter, a Puma, to replace the Piaggio, an aging amphibian which Alexander had been saying for months was a death-trap. Vacationing in Mexico with her children, Francesca and Lorne, Fiona heard the optimism and trust in Alexander's voice with misgiving. His father was as treacherous as the tides, Alexander had told her after Onassis had reneged on a deal Alexander had made him put in writing. "My word means nothing, whether I write it or say it," Onassis had laughed at his son's naiveté.[20] Why did Alexander believe that he could trust him now? But it was a question Thyssen did not want to pursue on the telephone, for she did not trust telephone, and she changed the subject.

Later that night, Onassis called Georgakis and told him that he had finally straightened things out with his son. In a couple of years he would be seventy (as usual, he erred on the side of subtraction: he was a week short of his seventy-third birthday); it was time they buried the past, he said.

It was almost five years since Bobby Kennedy's assassination. With Hamshari disposed of, Alexander appeased, and Jackie on her way out, things finally seemed to be going Onassis's way again.

On January 22, 1973, shortly before 3:15 P.M., aircraft SX-BDC Piaggio 136 of Olympic Airways reached taxiway F of Athens International to hold for takeoff. Alexander Onassis sat in the right-hand chair; Donald McCusker was on his left. Donald McGregor sat in the middle passenger seat behind them. McCusker had just arrived from Ohio, to replace McGregor as the regular Onassis pilot. However, McCusker had never flown a Piaggio, and Alexander had devised a plan to speed him through the familiarization flights. McCusker would be treated as a client; Alexander would check out his flying skills, then rent the plane to him. Accompanied by McGregor, McCusker would then "fly furiously" for a week to get his hours up before the plane was shipped to Miami to be sold.[21]

At 3:21 the Piaggio was cleared for takeoff. But within seconds of becoming airborne, the right wing dropped sharply and stayed down. There was no yaw or swing to indicate engine failure; there was no shuddering to suggest a stall. The plane just seemed to lose its balance and plunged to the ground.

Onassis was about to leave for lunch at P.J. Clarke's in New York when he heard the news. Tina, to whom Alexander had refused to speak since her marriage to Niarchos, was told while dressing for a dinner party at the Palace Hotel in St. Moritz. Christina heard about it on a car radio in Brazil. It was 6:30 in London, and Fiona Thyssen was preparing to go to her brother Richard's wedding dinner party when she saw a picture of Alexander appear on the television news, followed by a shot of the wrecked plane on the Athens runway. The last scheduled flights to Athens that evening had already left, and she spent the next three hours on the phone calling every person she knew who had a private plane (nobody from the Onassis office had called her or told her that Onassis had hired a British Airways Trident to fly the English neurosurgeon Alan Richardson to Athens). "All I knew was that Alexander was alive, and I was going to look after him; he was going to be all right," Thyssen told me.[22]

She arrived in Athens at six o'clock the following morning. Alexander was on a life-support system, following an operation to remove blood clots and relieve pressure on his brain, and he looked remarkably undamaged. (McCusker and McGregor had also been seriously but not critically injured.) "It's weird the things you remember," Thyssen would later recall. "Alexander was always so proud of his new nose"—she had persuaded him to have rhinoplasty a few months earlier—"and I thought: Thank God his nose is all right." Part of his hair above the right temple had been shaved, but except for a few scratches on his hands ("as if he had fallen on gravel," Thyssen thought), he looked fine.[23]

It seemed inconceivable when Alan Richardson confirmed the Greek surgeon's view that Alexander had suffered irrecoverable brain damage. Onassis summoned a third neurosurgeon from Boston. Fiona would recall the silence of the stopped lives of people waiting for someone to give them hope as they gathered in a small room across from the intensive care unit. She felt grateful when Jackie came and sat by her side. She knew that Jackie—who was also an outsider amid the family's tribal despair—would understand her feelings better than anyone else at that moment.

Jackie took her hand and said in her whispery voice, "Fiona, I have to talk to you." She knew that Alexander told her everything, she said; she also knew that Onassis had discussed their forthcoming divorce with his son. *Could Fiona tell her what figure Onassis had in mind for her settlement?* [24]

Expecting and dreading sympathy ("sympathy, when you're that vulnerable, you don't need it"), Thyssen was stunned by the question. Yet it was exactly the kind of distraction she needed at that moment, and as surreal as it seemed in retrospect she calmly told her that it was a question Jackie should address to her husband. Jackie agreed, and left Thyssen to her grief.

At one o'clock in the afternoon of January 23, the Boston specialist confirmed the English and Greek surgeons' opinion: Alexander had suffered general contusion and edema of the brain matter; the right temporal lobe had been reduced to pulp, and the right frontal fossa severely fractured. Only the machines were keeping him technically alive. Onassis told the doctors to wait until Christina arrived from Brazil to say goodbye to her brother—"then let us torture him no more." [25]

It was the only thing he had ever done in his life as a father that Thyssen "respected on any level at all." [26]

Christina arrived later that afternoon and sat with her brother for fifteen minutes.

At 6:55 P.M., the doctors switched off the machine.

The death of a son is a terrible blow to the psyche of a loving father, but it was far worse than that for Onassis. Filled with remorse for the way he had treated his son in his short, unhappy lifetime, already convinced that his death was no accident, but unable to decide which of his enemies to blame, Onassis was distraught. Costas Haritakis recalls going to the house on the evening Alexander died. "I kneeled down in front of him and grasped his knees. I said, 'Aristo, do you know how many millions in the world at this moment are crossing themselves and saying, *Thank God I am not Onassis?* Do you understand how many are praying that God will give you strength to face this terrible thing?' He said, *'They have killed my child.'* Those were his words. *'They have killed my child.'"* [27]

Who had killed his child? Who were *they?* Who would want to kill such a lovely boy? But Haritakis said simply, "Aristo, Alexandros is amongst the angels." [28]

Thirty minutes after the life-support system was switched off, Johnny Meyer was on his way to Monte Carlo to "clean up" Alexander's apartment. Onassis, he said, "didn't want to leave anything around that his enemies could use against him." [29] What he meant, of course, was what Georgakis would later call "the Onassis Family Jewels" *—the evidence of the part Alexander believed his father had played in Bobby Kennedy's murder. Whether this evidence comprised more than the Sirhan photostats is unclear, although Fiona is convinced that Alexander "knew far more than he told me. Looking back, I don't believe it would have been only the Sirhan material he put in the safe deposit box," she told me.

Onassis had hired Meyer partly for his ability to handle the press, and his good company, but mostly, one suspects, for Meyer's loyalty and trust qualities the American had shown while serving Howard Hughes, a man who had had as much to hide as Onassis—and for Meyer's willingness to do whatever it was his master wanted done. Nevertheless, although he probably believed that he had been sent simply to dispose of some smoke, a chore he had done for Onassis (and Hughes) many times, Meyer did not feel good about what he was doing in a dead man's home in Monaco. He spent three days going through the apartment, but unable to bring himself to use Alexander's bed, he slept on a sofa. "The dead kid was in that apartment with me," he told Brian Wells. [30]

---

* A reference to a secret report revealing the CIA's involvement in assassination plots against foreign leaders which had acquired the nom de scandale of "the Family Jewels."

He found the tapes of Onassis's accusatory, drunken, and abusive calls to Alexander hidden in a cavity at the back of a linen cupboard[31]; he also found several bugs, and evidence of wiretaps.* He did not, however, as far as we know, find the keys to the Family Jewels—which were locked, Alexander had told his father, in a safe deposit box in Paris.

Convinced that Alexander had been murdered to punish him—"my enemies would rather see me suffer than kill me," he told Georgakis—Onassis was too upset to attend his son's funeral on Skorpios, although he would later sit alone for hours at the graveside at night. "I would get so upset watching him come back to the house, eating alone two sardines and drinking a whisky at one o'clock in the morning," his housekeeper Georgia Veta recalled of the melancholy scene.[32]

But Alexander's death caused anger as well as sadness among those who suspected that he had died for the same reason Onassis's pilots Dimitris and Giorgio Kouris had been killed eleven months earlier—to punish Onassis. At the Glyfada villa, the day after the funeral on Skorpios, Fiona Thyssen overheard Gratsos and Georgakis discussing the tragedy in terms that horrified her. "It was as if finally they felt Ari had been paid back for all the terrible things he had done," she remembered the chilling sense of *schadenfreude* in these two old friends. "That freaked me: the absolute hatred of this man by two people who couldn't have been closer to him. He didn't blow his nose without one of them being there with the Kleenex, and they *despised* him. I thought, 'God, this man's worse than even I thought he was.'"

---

* Fiona Thyssen had warned Alexander that Niarchos, as well as his father, was tapping his telephone. Her suspicions were aroused when shortly after she sent Alexander a cable from Africa, where she was on safari with Stavros and Eugenie Niarchos, Stavros made a comment that revealed he knew exactly what she had written. "He could not possibly have intercepted the cable at my end; he could only have acquired the information from Monaco," Thyssen told me. It was the practice in the 1970s to read cable messages to recipients on the telephone, confirming it later with hard copy.

TWENTY-SEVEN

DOPE IS MONEY

It is only when the rich are sick that they feel the impotence of wealth.

—CHARLES CALEB COLTON, 1780–1832

Within hours of its completion on the morning of April 20, 1973, Yannis Georgakis handed Onassis a copy of the secret Greek Air Force accident investigation report on the crash: thirty-seven pages explaining why a plane fell out of the sky on a perfect winter afternoon.

Onassis held the folder in his hands as if it were something holy. He seemed reluctant to open it, as if afraid of what he might discover. He read the first page, then read it again, trying to absorb the technical detail of an unbearable loss. After a while, he tossed it back across the desk. "What's the point, Yannis. Nothing's going to bring back my son."[1]

The report concluded that the crash had happened because the aileron cables had been reversed during the installation of a new control column: Thus, when Alexander attempted to turn left on takeoff, as instructed by air traffic control, the plane would have banked to the

right; the harder he pulled the stick to the left, the steeper the plane
would bank to the right.

The cables had been reversed sometime between the evening of
November 15, when an Olympic engineer removed the Piaggio's control
column, and its fatal flight on the afternoon of January 22. A new column
had been installed by a second Olympic engineer on November 25. On
January 18, Greek CAA and Olympic inspectors approved the plane's air-
worthiness, subject to the formality of a test flight (which Alexander had
decided to combine with the flight to check out McCusker).

There were only two possible explanations for the reversed cables:
Either an Olympic engineer had seriously bungled when he installed the
new control system, and both the Olympic inspectors and the Greek
CAA had missed the fatal error; or Olympic security was so lax that
someone had been able to get into the hangar and sabotage the plane
between the inspection on January 18 and the fatal flight four days later.

The Olympic engineers challenged the Air Force findings. They
claimed that the color markings which are normally painted on connections before they are uncoupled (to establish what was joined to what at
the time of an accident) were, in fact, applied *after* the disconnection—
throwing doubt on the reversed cable theory. A second investigation—
conducted by Alan Hunter, one of the most experienced air crash
detectives in England was secretly commissioned by the airline. But
Hunter simply confirmed the findings of the Air Force report, conclud-
ing that there was "incontrovertible evidence that the aileron controls
had indeed been reversed."[2] But it didn't convince Onassis, who offered
one million dollars for information that would lead to the conviction of
the killer or killers.

Many accused Onassis of paranoia, but Meyer's discovery that
Alexander's apartment had been bugged by others, as well as his father—
a fact that was kept secret—lent some credence to Onassis's suspicions.
The Air Force had argued that since Alexander had not decided to check
out the new pilot until the evening before the accident, there was no
time for anyone to "undertake the lengthy job of clandestine sabotage by
reversing the aileron connections."[3]

Meyer's discovery of the bugging demolished this theory. In fact, Alexander had made his plans at least forty-eight hours earlier. He talked to Fiona Thyssen about them on the evening of Thursday, January 18, when he called her in Morges and told her that he would not be able to attend her brother's wedding in London because he planned to fly.[4] Therefore, whoever bugged his telephone would have had plenty of time to switch the aileron cables for the test flight four days later.

Nevertheless, what had begun as a personal tragedy for Onassis had become a corporate crisis for Olympic Airways, who were cast under a pall of suspicion that their engineers had bungled the repair. Despite sympathy for Onassis's loss, executives were losing patience with his obsession that his son had been murdered. Ari wanted justice, but the last thing the airline wanted was the publicity of a trial. It was a bad time for the airline business; the Palestinian terrorist campaign was at its peak, and hard questions were being asked about airline security. In an unguarded remark that made the rounds at the time, a senior Olympic executive said that he didn't give a damn who killed Alexander, he didn't care whether his skull was crushed with a hammer,* whether the CIA, the colonels, Niarchos, or a bunch of Palestinians did it, they couldn't run an airline as if it were an international conspiracy: "He's your friend. Shut him up, or he's going to talk this company right out of the airline business and into the toilet," he told Georgakis. But shutting Onassis up was easier said than done. "In moments of tragedy," Georgakis said in an attempt to mitigate Onassis's behavior, "a volatile character swings into unforeseen dimensions of illogic."[5]

Onassis meanwhile continued to entertain old friends aboard the *Christina;* people still came to stay on the island; he still drank his favorite Black Label scotch and sang his favorite Greek ballads at night;

---

* This was one of the many rumors floating around at the time, and the favorite explanation for why Alexander was the only one on board to receive fatal injuries. McGregor had a compression fracture of the spine, concussion, and leg injuries; McCusker, less physically injured, was suffering from amnesia.

but nobody was swept along by him any more. Aboard the yacht, after others had gone to bed, he walked up and down the deck until it got light, "like he was afraid to sleep in the dark."[6] On the island, unable to sleep, he would sit on his haunches like a peasant beside Alexander's tomb.

Meanwhile, his tanker business was thriving. Not since Suez had tanker rates risen so rapidly; VLCCs and ULCCs* made profits of four million dollars on a single run from Kuwait to Europe, and with global oil consumption increasing by over 8 percent a year, he ordered four more two-hundred-thousand-ton tankers from Japan, and two ULCCs from France. But on October 6, 1973, the Arabs launched the Yom Kippur attack on Israel, and the market crashed. Within weeks, more than a third of Onassis's tonnage was idle; he was forced to cancel the ULCCs from France, at a loss of $12.5 million. Many believed that the Soviets were behind the Arab offensive, and Onassis turned on David Karr for not warning him.[†] "If you're so fucking tight with the Russians, why didn't you get wind of this?", he screamed at him on the telephone on the morning of the attack.[7] Onassis had made a fortune when Randolph Churchill warned him about the Anglo-French plans to seize the Suez Canal seventeen years earlier. "Karr's no Randolph," he told Sir John Russell.[8] But his rage, the Englishman suspected, was not because Karr was no Randolph, but because he knew that he was describing his own tragedy: *"Onassis was no Onassis any more."*

Onassis's health was deteriorating. He was drinking more heavily than ever, and holding it less well. In the men's room at the Crazy Horse Saloon in Paris, he put his penis on the saucer in which customers left coins for the attendant. "That says it all: sex and money, the secret of my success," he roared, inviting a paparazzo to take a picture. Yet despite all this, Onassis was determined to remain a mover in the game.

---

* Very large and ultralarge crude carriers of over four hundred thousand tons.

† The Arab attack had taken everybody by surprise; twenty-four hours earlier, the CIA had dismissed the military activity in Egypt as annual maneuvers. (*Memoirs of Richard Nixon*, p. 920. New York: Grosset & Dunlap, 1978.)

But when another Onassis scheme—an attempt to build an oil refinery at Durham Point, a wooded headland above Great Bay on the Atlantic coastline of New Hampshire—ran into heavy local opposition, it was evident that he had neither the stamina nor the appetite for a fight. His sister Artemis urged him to retire. "He had taken so much punishment I was so afraid for him. He had thrived on pressure when he was a young man, but he was not a young man anymore," she worried.[9]

He lurched from outbursts of grief and self-pity to scenes of frightening anger; the smallest thing could set him off, especially if it concerned his wife. One evening, when Jackie told Georgakis that the book she was reading claimed that Christ was the creation of the Apostles, and asked whether he thought it was possible that Socrates was the creation of Plato, Onassis attacked her with such scorn that she fled from the house, wearing only a thin silk peignoir. "It was an appalling night: hailstones the size of golf balls . . . you wouldn't let a dog go out in it. Yiannakopoulos and I tried to stop her," Georgakis remembered. "Yiannakopoulos said she could be killed." With his typical crassness, and habit of playing fast and loose with the truth, "Onassis said it would save him a fortune. 'All she's ever given me is the Kennedy clap!' he said."[10]

It was not the first time Georgakis had heard Onassis complain that Jackie had a venereal condition inherited from her first husband.* But since Ari was still sleeping with other women—often women whose professional cries of ecstasy he still preferred to the real thing—to blame Jackie for giving him a disease was unfair, as well as cruel.

How much humiliation was Jackie prepared to take from him? Emmet Whitlock recalled an incident at a dinner party in New York: "I was talking

---

* According to his biographer Nigel Hamilton, John F. Kennedy had been treated since 1940 for a series of venereal diseases including a sexually transmitted bacterial disease called postgonococcal urethritis, usually contracted along with gonorrhea, "which not even penicillin, once available, could cure." (*JFK: Reckless Youth*, p. 342. New York: Random House, 1992.) In his 1997 book *The Dark Side of Camelot*, p. 232 (New York: Little, Brown & Co.), Seymour Hersh reports that Kennedy's venereal disease is known today as a chlamydial infection, which is "easily transmitted to women, and creates special risks for them."

with Ari and André Meyer, the head of Lazard Frères, and Jackie tried to join the conversation. Ari put his elbow up and just pushed her right out of the way. Oh yes, oh yes he did. He couldn't stand her. He couldn't stand her. He couldn't stand the sight of her," Whitlock said.[11]

Some felt that Jackie had never recovered from the culture shock of marrying Onassis, but had taught herself to appear oblivious to his boorish treatment of her. Why didn't she simply leave? Was she, as one friend cruelly suggested, "holding out for her widow's dues"?

It was certainly evident that Onassis was not a well man; he was, in fact, convinced he was dying. Not quickly, the way he had expected to die, but slowly, muscle by muscle, regret by regret, as if, in Georgakis's words, "a terrible retribution was at work."[12]

Costa Gratsos was shocked at Ari's appearance when they met in New York a couple of months after Alexander's funeral; Ari looked like "a besieged fortress crumbling from within," Gratsos told friends, and it was partly to take Onassis's mind off his grief that he ran their Haiti scheme past him again. Gratsos had never given up hope that they could turn that "sorry-ass banana plantation" into a Caribbean Monaco. Although Papa Doc had died in 1971, the infrastructure of corruption was as inviting as ever. The new President-for-Life, Jean-Claude ("Baby Doc") Duvalier, an obese playboy who had inherited his father's greed, but little of his ability, would be an easier man to do business with than the wily Papa Doc, and Onassis agreed to take another look at the plans they had devised a decade earlier.

Meanwhile, in spite of the promise he had made to Alexander to end his marriage to Jackie (a sacred vow in the circumstances, thought Georgakis), Onassis claimed that the Yom Kippur war had cost him too much to get into an expensive divorce battle too soon. He told Rico Zermeno that Jackie wanted $20 million. "I said, will you pay it? He said, 'My dear Rico, when you've been sleeping with a woman for ten years, she has to be very stupid indeed not to know at least one thing that could hang you,' " Zermeno recalled Onassis's reply.[13]

It was at about this time that Georgakis had breakfast with David Karr at the Travellers' Club in Paris. They were not close friends, but they shared what Karr called the "mad addiction"[14] of working for Onassis, and they occasionally met to swap war stories. When the subject of Haiti came up, Georgakis said that he didn't think it would happen; it would cost too much, and in the prevailing financial climate even Onassis might find it difficult to raise money for such an ambitious scheme.

"Finance will be no problem," Karr said, according to Georgakis. Karr, who appeared wary of directly involving himself in the matter, explained that he had introduced Onassis to an American banker and lawyer who would find "all the money Ari needs to pay for Haiti—and take care of Jackie's alimony, too."[15]

The man to whom he had introduced Onassis was Paul E. Helliwell. A paralegal in his Miami law firm, Helliwell, Melrose and DeWolf, remembers that Onassis and another Greek (Costa Gratsos) came to their office on Brickell Avenue late one night in autumn 1973. After-hours appointments were not unusual at HMD. "We had many famous clients who wanted to keep their dealings with Helliwell private," she said, and meetings were scheduled at all hours.[16]

"Paul loved intrigue," confirmed Helliwell's assistant Bill Losner. "He lived on the knife's edge. You wouldn't believe the payoffs; the celebrities I saw with crime figures when I worked for him you wouldn't believe!"[17]

But what stayed with the paralegal was Onassis's appearance. "He looked like shit. I thought, How can Jackie go to bed with this *old* guy? He looked like ninety years old." Her first guess was that he "wanted to launder some of his millions," but when another Helliwell client—Mitchell Livingston WerBell III—arrived shortly afterwards, she guessed that the meeting was about "something a little heavier" than money laundering.[18]

Although we shall never know precisely what was said that night on Brickell Avenue,* it is possible to surmise the parameters of the deal that must have been made from the events that followed—and the natures of the two Americans Onassis and Gratsos had gone to see.

---

* According to my source it was a "no paper" meeting, which meant that no record was kept, not even a note in the appointments diary.

Chief of the Office of Strategic Services in China in World War II, Paul E. Helliwell had famously traded information for opium.[19] After the war his Sea Supply Corporation, a CIA front, ran guns and dope to anticommunist Laotian and Thai mercenaries, who were key players in the Mafia's heroin trade.[20] In the late 1950s, he moved to Miami and the Bahamas, where his Castle Bank (reputedly set up to channel CIA funds for operations against Castro[21]) had become a Caribbean bridge "between the poppy fields of Thailand and organized crime in the United States."[22]

A friend of Helliwell's since their OSS days together in Indochina, Mitchell WerBell was not only an international arms dealer but also, according to the Washington intelligence reporter Jim Hougan, a preeminent specialist "in the crafts of assassination and the freelance *coup d'etat*."[23]

Although his role in subsequent events would remain enigmatic, WerBell's presence would give some credence to Karr's later assertion that whatever else Onassis discussed with Helliwell that night, Karr himself had merely arranged a meeting to raise cash to mount a private *coup d'etat* in Haiti—and pay off Jackie.*

The first indication that a major new heroin trafficker had arrived on the scene appeared in Houston, Texas, around the beginning of 1974. "It was high quality stuff, more potent, more deadly than Turkish smack, and there was a hell of a lot of it coming in," former Drug Enforcement Administration contract agent Basil "Beau" Abbott remembers.[24]

Because of his special knowledge of foreign smuggling routes, and enthusiasm for covert operations, Abbott, who had a reputation as "an intelligent, resourceful, and courageous"[25] agent, was ordered to try to infiltrate the new supply line. Going under deep-cover in Houston, the busiest inland port in the United States, Abbott set up a motorcycle

---

* Georgakis, who was inclined to believe Karr in this matter, said that Karr had also told him that he had been referred to Helliwell by Washington lawyer James Rowe, the foreign agent for the Haitian American Sugar Company of Port-au-Prince, and the lobbyist Tommy "the Cork" Corcoran, who, at United Fruit's behest had, according to Penny Lernoux, "helped trigger" the CIA's 1954 overthrow of the Arbenz government in Guatemala.

import-export business close to the port. Developing a sideline selling heroin to bike gangs, he quickly became one of the busiest dealers in the city, and a man with whom the suppliers wanted to do business.

"I really did a bang-up job. I found out which ships the stuff was coming in on; sometimes I even helped unload the stuff," Abbot told me. "The case was just building and building; I was getting a lot of bouquets. It looked as if we were going to nail the big guys this time." But at this point, the CIA intervened. "They said, 'That's enough, you son of a bitch, we're pulling the plug,' " Abbott recalled of the curt order that came from Washington.[26]

Although his line of work had inclined him to be philosophical about most things, he could not hide his anger and frustration. His case officer, Ron Gospadarek, told him he had ruffled feathers in Washington, and should let it go. Abbott did as he was told, but he couldn't just let it go. It piqued his curiosity. In his experience, cases were terminated only when "somebody very big, very rich, or wired to important people" was being protected.[27] And if somebody very big, very rich, or wired to important people was being protected, he at least wanted to know his name. He called in a favor from an old CIA friend, and a few weeks later got a call made from a pay phone in Washington.

"He gave me two names," Abbott recalled. *"Ari Onassis and his partner, somebody called Costa"* (written down as "Kostas" in the notes Abbott made at the time). Like his DEA controller, Ron Gospadarek, the CIA man advised Abbott to forget about it now that he had satisfied his curiosity. "Remember who Onassis is married to," he said.[28]

It was not a moral issue for Abbott. He knew how the world worked; he understood that sometimes compromises had to be made and favors exchanged; maybe the CIA even had something that might have been compromised by the DEA operation. But he also knew that a heroin trafficking operation on such a scale could not have continued without the complicity of people with juice in Washington*—people like Paul

---

* This cynical view was endorsed by an expert witness at Senator Henry Jackson's hearings on the narcotic rackets in 1975, who testified that without the right contacts, without the complicity and cooperation of law enforcement authorities and criminal justice

Helliwell and Mitchell WerBell, for example*—and what really pissed him off was the possibility that the Houston operation was aborted simply to save Jackie Onassis's blushes; or because somebody figured the family connection might hurt Ted Kennedy's expected bid for the Democratic presidential nomination in 1976.

But for whatever reason, "the case was crushed like an overripe mango," says an agent who had worked with Abbott in Panama and Belize (where Abbott had been commended for risking his life to thwart a plot to murder four DEA colleagues). This agent believes it was the cynicism with which the case was terminated that led to Abbott's own downfall (he was sentenced in 1990 to five years imprisonment with fifteen years parole for conspiring to smuggle drugs himself).[29] "There is much more to Mr. Abbott's story than meets the eye or is recorded in official documents," Abbott's former DEA supervisor Bill Coller admitted when he appealed against his conviction in a sworn affidavit in 1997.[30]

Apparently realizing that it was too dangerous to continue, Onassis ended his brief but enriching adventure in the narcotics trafficking business at about the same time that Basil Abbott's operation was abruptly closed down. "One quote sticks in my head," says Abbott now. "When I asked why the Houston operation had been turned off just as we were

---

personnel at all levels of government, "heroin trafficking could not continue to exist on any significant scale." (Chip Berlet, "Inside the DEA," *High Times,* Dec/Jan 1976.)

* One of WerBell's closest friends was former CIA comrade, Colonel Lucien "Lou" Conein, chief of the DEA's newly formed Special Operations Branch. With no time for the DEA's "buy and bust" operations, which were catching only small-time dealers, Conein aimed to resolve America's drug problem by what one of his associates described as "a process of elimination," and George Crile in the *Washington Post* ("A Soldier's Drug War," June 13, 1976) called "an assassination programme." The man from whom Conein proposed to acquire the weaponry for his "assassination bureau" was WerBell, with whom he shared his DEA safe-house in Washington, D.C. Conein's "assassination bureau" was terminated in January 1975 after Senator Lowell Weicker (R-Conn) revealed that Conein had been negotiating to buy (from a WerBell company) a consignment of assassination weapons, whose "only conceivable use was for anonymous murder." (*Washington Post,* June 13, 1976.)

about to land Onassis, a very senior DEA guy told me, 'Dope is money and money is power.' "[31]

Shortly before Yannis Georgakis died in 1993, he gave me the name of a man he said I should talk to in Palm Beach: William Carter. Carter had known Onassis since his arrival in New York from London in the 1940s. Carter had been part of the New York social milieu that Onassis admired so much, and set out to gatecrash. Carter became a regular weekend guest at Onassis's cottage on Center Island on the north shore of Long Island, which Onassis shared with his Norwegian mistress Ingeborg Dedichen. "Ari was a good sport, a lot of fun. We regularly did the rounds together—the Stork Club, "21", Elmo's [El Morocco], the Colony, and a place on East Fiftieth Street he liked, called Versailles. He must have been fifteen, twenty years older than me but he'd still be going strong after I was tucked up in the feathers. The energy that guy had! When I told him I was going home to England to join the army, he said, 'Who the hell am I going to cut up with now?' "

Georgakis told me to ask Carter about Onassis's excursion into drug-running. Carter, he said, had "his finger on the pulse of everything that moves" in Florida. In the spring of 1995, I visited Carter at his home in West Palm Beach. A former Vatican ambassador to the United Nations, a Knight of Malta, and Palm Beach socialite, Carter was plainly in a mood to talk as he stretched out on a chaise longue in a living room overlooking a walled garden filled with roses and bougainvilleas. Books, little tables filled with mementos and bric-a-brac filled the room; cut-glass decanters of Madeira wine and spirits were on a butler's table by his side. He wore khakis, Gucci loafers, no socks, and a white, short-sleeved polo shirt. Tall, thin, with the look of a ravaged English aristocrat (he was born in Derby, England, in 1918), he said cheerfully, "You've caught me just in time. I fancy I don't have long for this world." He had, he said, prostate cancer. "What can I tell you?" For an afternoon he reminisced, he wandered, but this is the story he eventually told me:

In the autumn of 1973, his old friend Johnny Meyer invited him to dinner at the Colony hotel in Palm Beach. They had known each other since

1946; shared adventures had deepened their friendship; they had invested in property in West Hollywood together, even shared a mistress, whom they set up in an apartment in Beverly Hills. "Johnny said, 'Bill, if we can share this beautiful woman and not want to kill each other, we can share anything.' " What they shared best were secrets. "We were like brothers; we could tell each other anything, and know it would go no further."

Over dinner that autumn, Meyer told him that although Onassis was still seriously rich, he had a temporary liquidity problem. "He said, 'Ari's got these big plans to develop Haiti, and he wants to dump Jackie, and he wants to do them both in a hurry. It's going to cost an awful lot of dough, which Ari doesn't have to hand right now.' I said, 'Well, I don't know a lot about Haiti, but I've known Jackie since she was a Bouvier— I went to her debut at the Clambake Club in Newport in 1947—and one thing I know about Jackie is that she knows her price and won't chisel down.' Johnny agreed. He said that was why Onassis was in Miami right now talking to Paul Helliwell at Castle Bank. Well, Helliwell and Castle Bank meant only one thing to me: drug trafficking.

"I asked Johnny whether Ari was thinking about doing something screwy," Carter continued. "He said, 'Ari's in Miami; I'm here with you; what do I know?' " But Meyer didn't look happy, and Carter says he didn't pursue the matter, although he had his suspicions. In fact, Meyer had once told Carter that Ari had confessed to him that in the thirties he had "made a little money" trafficking drugs between Argentina and Europe; he quit only when the Italians became suspicious after they discovered an insurance scam he was running, and jailed one of his cousins.[32]

"Ari was simply going back to the business that got him started. You don't think he really made his first fortune shipping tobacco, do you? *He* started out running dope, and he finished up running dope. I don't know that he ever admitted them, [the rumors of his drug-running days that Carter claimed followed Onassis to New York in the 1940s], but I don't recall him ever denying them either. He was a ruthless adventurer, but I was fond of him." *

---

* Arthur William Carter died on October 30, 1995.

Whatever service WerBell had performed for Onassis—whether it was to help him steer through the shoals of heroin running, or to assist his proposed *coup d'etat* in Haiti*—Onassis was clearly satisfied with the results. In a touching exchange of gifts, he gave WerBell introductions to the politicians and officials responsible for re-equipping the Greek mili-

---

* After Onassis's death, Gratsos invested his share of the heroin profits in a final bid to grab the prize that had eluded him and Onassis for so long: Haiti. In 1978, again enlisting the services of Mitchell WerBell, and with an exiled Haitian banker and former Papa Doc bagman, Clemard Charles, he resurrected the plans for a *coup d'etat*. Charles claimed that he had an army of one hundred and fifty men in the Haiti jungle waiting to join a party of former Special Forces troops to parachute into Port-au-Prince to take the palace and arrest Baby Doc. Bill Jordan—who was the LAPD night watch commander at Rampart detectives the night Robert Kennedy was shot, and the first officer to interview Sirhan Sirhan—told me that his security company had been hired to handle the policing of the island after the takeover. His team went through a training course at WerBell's camp in Powder Springs. "It was a hell of a group. All ex-Special Forces people. I had one lieutenant-colonel and half a dozen bird colonels. "We were going to wait in the Dominican Republic and when they put it out that they had taken over, we were going to fly straight into Port-au-Prince," Jordan recalled. "I was going to bring in some police administrators, some really sharp guys, and set up a training academy for them and make real honest-to-god policemen out of them. I don't think Gratsos gave a damn about the Haitians. He had a whole program of investments worked out, big development plans, tax-free deals for international companies, the promise of a cheap work force. "I got really wrapped up in that thing. The poor Haitians are the most oppressed people in the world. They're not bright, they've got no education. Helping to get rid of a shmuck like Baby Doc—I had no trouble justifying that. It wasn't as if we were going in to rape some innocent country and overthrow good people. That was the saving grace of the whole thing. It would have been good for the people. Jesus, there was no way you could make it worse for them." But the closer Jordan looked at the plan the less he liked it. "Clemard Charles was the problem. First, he gave me this short list. These are very bad people, they must be eliminated,' he said. Every time I saw him he'd hand me another list. It was beginning to look like the Haitian phone book. I was sticking these things in the round file basically. But I knew that if the deal had been right, Gratsos might have gone along with a short list. I also knew that once a guy like Clemard Charles got in you might be looking back and thinking what a wonderful president Papa Doc was and what a wonderful guy Baby Doc was compared with this butcher. Eventually, Jordan learned that Charles' claim that he had one hundred and fifty men waiting to join the invasion force was a pack of lies, and without at least the appearance of a popular uprising of Haitian citizens the invasion would have been a

tary after the fall of the colonels*; and WerBell sent him a prototype of a machine that analyzed voice modulations on the telephone, in order to determine whether a caller was lying.

Recalling how her father would call Jackie and question her about what she was up to when they were apart, Christina Onassis told friends that it was the most useful toy her father ever had.

---

public relations disaster as well as a bloodbath for the small invasion force. "I liked Gratsos, and I think he respected me. I knew he'd already spent a fortune on the operation but there was no way I could let him go ahead with a plan that would send men to their certain deaths. I told him to get out. It wasn't going anywhere. That was the end of it."

* How much business WerBell did in Greece is unclear, although the potential was considerable and Onassis's introductions worth their weight in gold. "They've got a funny situation over there," the arms dealer told friends when he returned from Athens in the fall of 1974. "The past regime received a lot of munitions aid, but they sold it to other countries. Now the Greek army is almost bereft of adequate weapons." (" 'We're Very Loyal Americans'—WerBell," *Atlantic Constitution,* October 14, 1974.)

TWENTY-EIGHT

GOD'S PUNISHMENT

Our repentance is not so much regret for
the evil we have done as a fear of what
may happen to us because of it.

—FRANCOIS DE LA ROCHEFOUCAULD, 1613–1680

For months Onassis had been experiencing bouts of extreme weakness; once or twice he had been unable to rise from his chair, or even open his eyes. In March 1974, his doctors diagnosed myasthenia gravis, a disorder of the body's autoimmune system. Since it usually hits men in their forties, Onassis jested that it was proof of what great shape his body was in! However, bloated by the cortisone injections prescribed to counter decreased adrenal function, his face belied his attempts to make light of his condition, and in more serious moments he told friends, "It's probably God's way of punishing me."

He had lost his only son; Omega was in ruins; his New England refinery project had collapsed. In Greece, Papadopoulos was in jail, awaiting trial on charges of high treason; and the new prime minister Constantine Karamanlis, pressed to deal severely with *junta* collaborators, refused to take his calls. In New York, despite a costly advertising campaign, only thirty-five of the two hundred and fifty apartments in his

Olympic Tower apartment block on Fifth Avenue had been sold. "Risk spoke to something abiding in Onassis's nature," said Georgakis, who had tried to dissuade him from the venture at a time of inflation and deepening recession.[1]

But if all this was indeed God's punishment, others would also get hurt in the collateral fallout.

The first disquieting signs in Christina's behavior appeared one morning a few months after Alexander's death in 1973, when she suddenly inquired after Alexander, as if his absence had only just struck her. "Where is my brother?" she interrupted a conversation with Harry Gatzionis, who had been charged with watching over her progress at Victory Carriers in New York,* where she was attempting to learn her father's business. Shocked, Gatzionis told her gently, "He's in heaven, he is gone. Now you must learn."[2]

Controlling her emotions had never been Christina's strong suit, and with Alexander no longer there to share the burden of the Onassis legacy, the strain was becoming evident. At the beginning of August she returned to London, and for twelve days did not leave her house in Reeves Mews, Mayfair. She refused to clean her teeth (an old childhood trick of rebellion and perhaps self-loathing) or even to get out of bed. And in the early hours of the 16th, her maid, Eleni Syros, realizing that Christina was slipping into a coma, summoned an ambulance.

At the Middlesex Hospital, Christina's stomach was pumped, and she spent twenty-four hours in ICU before being moved to a public ward under the name C. Danai. Because of his own failing health, Onassis was kept in the dark. Her mother, however, rushed back from the South of France, and for forty-eight hours did not leave Christina's bedside.

Maternal devotion was a gift few associated with Tina, and her vigil perplexed those who knew her well. But fearful of what her daughter might reveal in her disorientated state, was Tina's reluctance to budge from Christina's bedside closer to surveillance than a vigil?

---

* The company set up as part of the settlement of the 1950s criminal litigation brought against Onassis by the U.S. Justice Department, and was run by Constantine Gratsos.

Later, she told friends that Christina had tried to kill herself because her affair with the shipping heir Peter Goulandris, to whom she had been unofficially engaged on a number of occasions, was over. "Peter says he can never marry me," Tina claimed her daughter had told her. "I don't want to live."[3] But this explanation did not ring true. Their relationship had always been more a question of pragmatism than passion, and according to one of Goulandris's friends, Christina "treated him like shit."[4]

It had been a bad time for Tina, too. Her marriage to Niarchos had not brought her the satisfaction she had expected. Her husband continued to have his mistresses ("he was a womanizer, and what can we do about that?" sighed one woman friend[5]) and Tina consoled herself with champagne and lovers of her own. Her beauty was fading, partly because of her habitual reliance on barbiturates ("she was so doped up in the morning that she couldn't speak," said another friend[6]), and partly because of her drinking.

Friends were already fearing for her sanity, and some for her life.

TWENTY-NINE

A GOOD PLACE
FOR CONFESSIONS

It is wretched to be found out.

—HORACE

"F ATE," Aristotle Onassis would say when something had gone badly wrong in his life (and friends had heard him say it often since the former First Lady became the second Mrs. Onassis), "fate happens." And had not the world heavyweight-boxing champion George Foreman cut his eye in training in the autumn of 1974, the truth about Onassis's marriage to Jackie Kennedy might have remained Onassis's last and darkest secret of all. But a passionate reader of the Greek classics, he would have known as he sojourned on Skorpios that when fate happens, it is almost always determined someplace else.

In Paris, Hélène Gaillet sat on the terrace of Fouquet's restaurant on the Champs-Elysées and pondered the problems she faced because of the damage to the champion's eye, which had caused a postponement of his title defense against Muhammad Ali in Kinshasa, Zaire. A New York photographer covering the contest for the Gamma-Liaison agency, Gaillet had arrived in France en route to Africa on the morning the postponement was announced in the Paris newspapers. The arithmetic looked bad no matter which way she added it up: To hang around Paris for a month might be more appealing than schlepping back to New York, but it would cut deeply into her profit margins on the assignment. The following day, on her way to the PanAm office to book a flight back to New York, she had an idea.

Her lover was Felix Rohatyn, an investment banker and Wall Street superstar.* Although he had always kept a low profile, Rohatyn was a powerful man with even more powerful friends, and the year before he had introduced Gaillet to Onassis at a New York dinner party. A slender woman of thirty-four, with expressive eyes the color of old pennies, high cheekbones, and a finely sculpted mouth, Gaillet had intelligence as well as *savoir vivre*. Born in France and able to converse comfortably in French, English, Spanish, and Italian, she was used to the company of successful men, and successful men were attracted to her. Onassis was no exception. But although he boasted that he approached every beautiful woman as a potential mistress, he did not make a pass at her, as he increasingly did at friends' wives and partners now that his marriage to Jackie had become unsatisfactory to him, and his despair deepened.

But Gaillet, who had gone to live in America as a child shortly after World War II, remained a Parisienne in her soul, and knew enough about men like Onassis to understand that his restraint was probably as much a seduction ploy as a measure of the respect he had for her lover. She was not surprised, therefore, when shortly after their first meeting Onassis placed her next to him at a dinner he gave at the Coach House, one of his favorite New York restaurants. But as if implicitly laying down his rules—Gaillet had heard the rumors of his sadism with women—he ignored her for most of the evening. "Ari and Felix were talking business; when men like that discuss business their concentration is total. I just listened. I was a good listener; sometimes you can be helpful just by saying nothing," she said, with the confidence of a woman who knows that however badly they may seem to behave she will always be a challenge to such men. Sure enough, at the end of the evening, Onassis lifted the

---

* Rohatyn would later make his name as the man who saved New York City from bankruptcy. In 1997, President Clinton appointed him U.S. ambassador to France.

palms of her hands to his lips and told her that next time she was in Paris and needed a place to stay, she must call him. "I still remember the gleam in his eye," she would later tell me, recalling the invitation.

So, stuck in Paris, she telephoned Rohatyn in New York and asked whether she should now make the call. Her lover told her to go ahead: Ari would love to hear from her, he said. But when she rang Onassis's Paris number, she was told that he was out of the country and not expected back for some time. She knew that he could go to earth better than anyone when hunted by reporters, mistresses, or, increasingly in recent times, his wife. Nevertheless, she left her name and telephone number at the hotel anyway, and fifteen minutes later he called her back.

"I told him about the delayed fight and my frustration, and he said in that growly voice of his, 'Do you want to come here? I'm on my island.' I said, 'Yes, I was hoping.' "

And so began the fateful chain of events that more than a quarter of a century ago had led her to Skorpios in the final autumn of Onassis's life.

I had last met Gaillet in New York in the fall of 1983, when I was writing my Onassis biography. She had been a valuable source: a social insider with a good ear for the nuances of gossip, and she was observant, witty, wise, and generous with her personal reflections. I knew that she had been discreet about some things in 1983—she would not reveal the name of her lover, Felix Rohatyn, who had introduced her to Onassis, for example; she insisted that she had not slept with the Greek tycoon on Skorpios; but she told me things that confirmed and clarified several important points. It was only after the publication of *Ari* that I discovered how significant some of her discretions had been. Shortly before his death in 1993, Yannis Georgakis, who knew where so many of the bodies were buried, and told me where to dig, urged me to talk to Gaillet again: "She's important. Get her to tell you the *whole* story of her visit to Skorpios," he said.

But I had lost touch with her after *Ari* was published. Her long affair with Felix Rohatyn over—and after a break-up which she admits emotionally destroyed her for several years—she had remarried in 1990. Living an entirely different and very private existence, moving in a new social environment, among friends and neighbors who knew little or nothing of her previous life, she hadn't been easy to track down. I finally caught up with her in the autumn of 2002.

Now in her early sixties, still strikingly attractive, with shoulder-length, caramel-blonde hair, and a figure kept lean and fit by a passion for sailing, Hélène Gaillet de Neergaard, as she was now called, wife of a successful, retired chemist, William Field de Neergaard, was at first reluctant to talk about those days when, with a couple of phone calls, she could arrange an invitation to a billionaire's private island. "When I lived at that level I accepted things, I knew things, and did things, that now frankly amaze me," she admitted when we met for lunch at a smart yacht club in Naples, a small resort town on Florida's Gulf Coast. But although I was questioning her on matters she said she had not thought about in a long time, she hadn't forgotten a thing: how Onassis had sent his limousine (along with a personal stewardess and *valet de pied*) to take her to Le Bourget, from where his private plane flew her to Athens; how after a brief stay at his home in Glyfada, with his sister Artemis and daughter Christina, waiting for the weather to clear, his helicopter had taken her to their rendezvous on Skorpios.

She had met Christina half a dozen times in New York, where the heiress had been sent after the death of her brother Alexander to "learn the ropes" of her father's tanker business. "She was very dark, very thin, a strong personality," Gaillet had told me in 1983. "I found her fascinating, although she had this underlying aura of . . . I don't want to say doom, but doom's close." Christina had in fact been in the care of Artemis and her physician husband, Dr. Theodore Garofalides, after a serious suicide bid in London.[1] Although Gaillet was unaware of this at the time, she thought there was "something about her that you sensed was getting out of control, something in her eyes."

The Glyfada house was the most modest and—situated beneath the flight path of Athens International Airport—least impressive of Onassis's many homes around the world. But it was the house in which the family felt most relaxed, and in some sense were most free, and where they entertained only their most trusted friends. The décor was modern, almost minimal, except for a few pieces of dark, sturdy furniture, a table of fretted wood inlaid with mother-of-pearl, and a divan covered with a Kurdish rug, that had been rescued from the family home in Smyrna* after they had fled to Greece following the Turkish massacre of 1922.

After dinner, Artemis excused herself—the Garofalides had an almost identical house next door—and the young women continued to chat. When Gaillet asked whether Jackie would be going to the island during her stay, Christina answered bitterly that one of the reasons her father was on Skorpios was to get as far away from his wife as possible. "She said something to the effect that 'the quicker he gets rid of her, the better,' " Gaillet recalled.[2] It was no secret that Christina could scarcely stand to be in the same room with her stepmother, but Gaillet was taken aback by the vehemence of her outburst. Although she suspected that the marriage was over for Onassis, she was amused when Christina dismissed it as a *mariage blanc.* "I said, 'Christina, you can't possibly believe that the marriage has never been consummated?' She said, 'Oh, I'm sure my father's had his money's worth in that respect, he'd have made sure of that, and I believe Jackie is very good at it, a whore in the bedroom and all that, but he grows tired of all his women sooner or later.' "[3]

"Christina particularly resented the fact that Jackie made no effort to comfort Onassis when things were getting tough for him," Gaillet added. "She was not the kind of woman who would go and sit and hold his hand and say, 'Oh my poor darling, let's go out and have some fun together, let's go to the opera, let's go to the theatre, let's travel somewhere'—Jackie loved all those things, but not with her husband."

Christina had become fiercely protective of her father since his health began to deteriorate after Alexander's death. Although Christina said that her father looked forward to Hélène's visit, he was doing poorly and looked lousy; so she made Gaillet promise not to take any pictures of him on the island. Onassis had some business problems and "it wouldn't be good if his enemies saw pictures of him in his present shape," Christina told her.[4]

---

* Now Izmir.

When the weather finally lifted, Gaillet was flown to Skorpios. The epitome of exclusivity, it made the Hamptons seem like Coney Island in August. Onassis had spoken proudly of the scorpion-shaped island's beauty many times, and she was not disappointed. "But oh, my dear, poor Ari . . . he looked as if he had about an hour to live," she said.

Although she was his only guest, he had arranged for her to sleep on his yacht, *Christina;* he would stay at the house. "Here we live by the calendar, not the clock," he explained the island's routine. The first evening they dined aboard the yacht. He had taken a nap in the afternoon; and shaved and refreshed, he looked almost like the Onassis she had first met in New York. But she suspected she was looking at a man saying "fuck you" to his own pain.[5]

"When I was young," he told her as if he had guessed her thoughts, and with a hint of his old earthy sensuality, "I was sure I would have a vigorous old age. Now I am not young anymore I can't be sure of anything, except that we are all born and we must all die. But right now, here with you, Hélène, I am very glad to be alive."[6]

The next day, he called and asked what she was doing. "What am I doing? I'm reading books, having a wonderful time." Gaillet told him. He suggested they go for a ride. "He drove me all over the island in his Jeep, showed me the orchards, the farm animals, the flowers. The island was like a medieval domain . . . it produced its own milk, bread, meat, figs; only the fresh water had to be shipped in, and stored in reservoirs." Finally, he showed her where his son was buried. Although his sense of loss was palpable, it wasn't a sad visit. "Ari talked as if he expected Alexander to join us any minute. He said, 'Alexander is just as living to me as you are. He comes to me often. Unfortunately, till I die I cannot go to him,' " recalled Gaillet, who also knew from a maid that at night he often stood at the door of Alexander's bedroom, crying for his son's forgiveness.*

In the following days, his energies temporarily revived by Gaillet's presence, life on the island followed the same relaxed pattern. Sometimes

---

* Once, when I called Onassis in Paris not long after Alexander's death, Christina answered the phone and told me that her father was on Skorpios. Was Jackie with him? I asked. "No, he's with Alexander," she said.

Onassis joined her for lunch aboard the *Christina,* other days he did not appear at all. She always let him set the agenda: Forecasting his moods was a mug's game. One afternoon they drove to the other side of the island. They stripped and went into the sea. But despite the temptations of being with a beautiful naked woman on a deserted beach, Onassis still did not hit on her. Although his illness, his age (he was a couple of months short of his seventy-fifth birthday, but claiming to be five years younger), and an unreliable libido may have had something to do with his propriety, Gaillet also felt that he "rather liked the idea that there was a woman who was off-limits," as I had written in *Ari.*

She was impressed by his aura of power. She saw, as many did not, that although he was a rich and ruthless operator he was not *just* a rich and ruthless operator. He could be manipulative, irascible, vengeful, a monstrous figure—he spared no one, not even members of his own family, if they threatened to get in his way—yet he could also be kind, sentimental, articulate, self-mocking, and extraordinarily generous. "He was endlessly fascinating," Gaillet told me. "I adored his mind. He was a man who had never played safe in his life. Nobody expired from ennui around Ari. How could you not become absorbed in a person like that? There were endless moments when I thought, 'I could really get involved with this incredible man.' I knew that he must have made it with a hundred thousand women in his life, probably a few boys too, but we connected on another level entirely. We were having fun. Why have sex?", she had told me in 1983. They had talked endlessly about life, relationships, about the failure of his marriage to Jackie, who was her friend, too. "We talked about everything; we really got into each other's heads. He talked to me about his life; he went from one part of his past to another—his childhood in Smyrna, the affairs he'd had in Buenos Aires in the twenties and thirties; his escapades in New York and Hollywood during World War II.

"Skorpios," she said, "was a good place for confessions."

But in the summer of 2003, at her home high above a cove on Long Island's North Shore—the Gold Coast—she made a confession of her own: Although she had not lied to me in 1983, she said, she had been economical with the truth. "When I knew you were coming to see me again I thought: 'Oh my god, am I ever going to tell this story, or not?' Because I am a proper person: I don't talk about things I've done. Then I thought, 'Well, you know what? I think this is the right time for me to tell this story. This is the right time.' "

She had always told me the truth, she began, sometimes not *all* the truth, but *only* the truth. "Felix knew that I was a wall of silence, no matter what. And I've been a wall of silence for nearly 30 years. But now that so many of those characters have departed or faded away—and I'm not getting any younger—I feel that the whole truth should now be told. So many people are into revising history, I believe that what really happened should be put down by an historian, so that a hundred years from now people will know the truth about at least one part of our history."

She started on a personal note: Onassis *had* hit on her—and she had eagerly responded—when they swam nude together in the sea. "We made love. It was so physical, an extraordinary moment. We both knew when the frolicking began in the water neither of us was going to resist. I was not going to say no and he was not going to stop. I don't remember a word passing between us. It was getting on toward late in the afternoon; I remember the sun going over the promontory throwing a shadow across the beach. Ari tossed a towel on the sand and pushed me down and made love to me again. He didn't force himself on me. Are you kidding? I was in seventh heaven. I was enthralled. But it was purely a momentary sexual passion. He never attempted to make love to me again; he never made another move toward me. It was an encounter with an extraordinary man at an extraordinary moment that could never be repeated. We both knew that. We knew that if he'd said to me at the villa, 'Come on, let's do it again,' it would never have been the same. Perhaps that is why we got along so well together: we had a profound understanding of each other. That is why we were such great friends. He hadn't asked me to the island to fuck me. He had no idea, and I had no idea, that that moment was going to happen. It wasn't premeditated, or planned, or calculated. It was simply a moment. I knew his wife, I was in love with Felix, but not for a second did I hesitate or feel guilt, and I imagine it was the same for Ari. What we did showed a lot of mutual trust. We both had a lot to lose. The fact that nearly thirty years later I'm talking about it for the first time to anyone proves that. Not even Ari and I talked about it afterward; we never mentioned it again.' "

The discovery that Hélène Gaillet was alone with Onassis on Skorpios came as a shock to Costa Gratsos, the most influential figure in Onassis's life, and his oldest and closest friend. They had met in the 1920s in Buenos Aires, where Onassis had made his first fortune, shipping Oriental tobacco from his father's business in Greece to produce his own cigarettes, including a brand he called Bis; an enterprise that ended abruptly when he was sued by the proprietors of a hugely popular brand, also called Bis. Onassis protested that he had never heard of the other Bis, which was as plausible as Frank Sinatra denying all knowledge of Jack Daniels; and he settled out of court. "Costa knows every crime I've ever committed," was how Onassis had introduced me to Gratsos at Maxim's in Paris in the autumn of 1967.

"Costa rang me from New York as soon as he heard that Ari was entertaining a woman on Skorpios," Yannis Georgakis later told me. Gratsos had never sounded so angry in the thirty years Georgakis had known him. "He said, 'Do you know who this woman is? How did she get there?' I had no idea. I told him to call Christina, but he didn't want to do that. Christina was supposed to be working for him, and their relationship was 'rather difficult,' " Georgakis recalled.

Georgakis liked to think of himself as Onassis's *eminence grise,* and he certainly had the qualifications for the role. He had entered the Athens University Law School at fifteen, graduated at twenty, and continued his studies at the universities of Munich, Heidelberg, and Leipzig, where he obtained his doctorate in criminal law in 1938. A genial man with an epicene taste in clothes, pouchy eyes, and crumbling cherubic looks, he had served as legal adviser to Archbishop Damaskinos, the Orthodox Primate of Greece. He then became head of Damaskinos's political office when, following the liberation of Greece after World War II, the archbishop was named Regent, pending a referendum on the restoration of the monarchy. Later, Georgakis was appointed under-secretary with a special mandate for the Ionian Island, and also served as the first governor-general of the Dodecanese Island, Italian possession ceded to Greece in 1947. More recently, he had become ambassador-at-large to the Arab oil-producing countries, an appointment that some believed he had sought more to further Onassis's interests than his own. But if his true motives were often inscrutable, there is little doubt that after a lifetime of fulfilling minor but expedient political posts, Georgakis had developed a taste for influence and intrigue rather than personal power.

"All evil geniuses need a smart mouthpiece," Onassis had told him at the start of their relationship, and Georgakis enjoyed telling stories of Onassis's skulduggery.[7] The first present Onassis gave him was a copy of Aeschylus's *Eumenides,* the story of how the so-called principles of vengeance were replaced by the rule of law in ancient Greece. It was to remind him that Onassis was not a civilized man: Once he understood that, Onassis had told him, they'd get along fine. When I once asked Georgakis why he associated himself with such an obvious rogue, he said that as a lawyer it was his duty to defend the guilty, adding puckishly: "The people wish this, the law permits it, and Ari pays me good money to ensure that the practice continues."

According to Count Flamburiari, a social acquaintance of Georgakis's, whose family is one of the oldest and most aristocratic in Greece, it was this pragmatic attitude that enabled Georgakis to know "what was under every stone before it was lifted."[8] And although he told one friend that he never intended to get any closer to Onassis than their professional relationship required, Georgakis had clearly strayed close enough to know exactly why Gaillet's presence on Skorpios in the autumn of 1974 had alarmed Costa Gratsos so deeply.

A few weeks earlier, booked into a New York hospital under the name of Mr. Phillips for further tests and treatment for his steadily worsening condition, Onassis had had what Georgakis described as "an episode."[9] It was a stressful and unhappy time for Onassis. The collapse of his plan to build a refinery in New Hampshire; Christina's attempt to commit suicide in London; and the Greek government's determination to take back his flagship airline had sent him into a downward spiral of despair.

The "episode" (probably a mild form of stroke, known as a transient ischemic attack) lasted perhaps less than a minute. Onassis, heavily medicated, had been sitting up asleep in an armchair when Gratsos arrived. Although the doctors spoke optimistically to Christina, who still believed that it would only be a matter of time before her father was well again—"anyway, well enough to divorce Jackie," she had told her first ex-husband, Joseph Bolker[10]—Gratsos had been given a more realistic, and far grimmer, prognosis.

Locked in his own thoughts, Gratsos did not at first pay attention when he heard Onassis begin to mumble. It was not until he realized that he was addressing somebody else in the room—somebody only Onassis apparently could see—that Gratsos began to make out what he was saying. And what he was saying was truly alarming.

"Costa telephoned me from the room. I was in Athens. It was the only time I'd ever heard real panic in his voice . . . he said that Ari was confessing everything," Georgakis told me. They had no idea whether it was the first, or one of many similar "episodes," or whether it was the result of the medication Onassis was on for the disease that was slowly, systematically breaking down his autoimmune system. But the risk of it happening again, when other people might hear, was all too clear. Georgakis asked Gratsos how long he thought Onassis had to live. The most pessimistic estimate he'd been given was six months, the most optimistic a year. Georgakis said that six months was a long time to expect to keep Onassis 'out of harm's way.' " "Nobody will ever keep Ari out of harm's way," Gratsos told him grimly.[11]

The rapid deterioration of Onassis's health was hushed up, but Gratsos and Georgakis's assumption that he would continue to undergo treatment in New York was jolted a few days later when he discharged himself from the hospital and flew to Skorpios, where soon after, he would receive Gaillet as a guest.

Still waiting for George Foreman's damaged eye to heal, Hélène Gaillet was happy on Skorpios. "I was having a wonderful time," she said in the summer of 2003, when she had decided to tell me the whole story of that visit. The only guest on the island, waited upon hand and foot by a retinue of servants, having one of the richest and most fascinating men in the world all to herself, life didn't get much better than that. She was flattered that Onassis felt so comfortable with her.

Johnny Meyer had learned of her presence on the island in a phone call from Onassis. "She's easy on the eye, and has a smart head on her shoulders," he told his public relations man and old drinking crony. Aware of Hélène's friendship with Jackie Onassis and the Kennedys, Meyer warned him to be discreet. Onassis said, "You don't have to worry about Gaillet. She's Rohatyn's girl . . . she knows how to keep her mouth shut." [12]

Nevertheless, alarm bells started to ring in Paris and New York. A couple of days later Gratsos and Georgakis, accompanied by Meyer, arrived on the island. They were closely followed by David Karr. Although these four men could hardly have been more different from one another as human beings, each had his own reasons for wanting to keep Onassis away from outsiders while he was prone to such alarming indiscretions.

David Karr's past association with the Communist *Daily Worker,* as well as the *Washington Post,* whispers of a KGB connection, his closeness to eminent investment bankers and ties to Washington politicians, had continued to make it difficult than ever to be sure whose side he was on, or what his real purpose was. According to his London business partner Ronnie Driver, when a lover was giving Christina Onassis a hard time, Karr told her, "I'll get rid of him for you if that's what you want, just say the word." Said Driver, "If David said he'd get rid of somebody, that's exactly what he meant: *he would get rid of them.*" [13] From the time he took up residence in Paris in 1967, Karr dealt directly with Onassis instead of through Gratsos or Georgakis; it was an intimacy that did not please the two Greeks, especially when Karr became Onassis's neighbor on avenue Foch.

Gaillet knew neither of the Greeks nor David Karr—whom she thought looked like the kind of man who "would always make sure he was left standing when the smoke cleared"—but she had met Meyer in the sixties at a weekend house party in the South of France when they were guests of Daniel Wildenstein, the millionaire French art dealer and racehorse breeder. "I remember I couldn't figure out how Johnny fitted in with Wildenstein's crowd. When I asked a friend about him, she said: 'Don't you know? Johnny's a *racolleur*.' It's an old French word that means a man who puts people together: finding women for men, arranging men for women. I guess there's another word for it, but *racolleur* sounds nicer," she told me.

The moment Meyer and his companions arrived on Skorpios (all of them incongruously wearing suits and ties, like visiting bankers) Hélène felt the atmosphere change. Were these the people, she wondered, who really ran things for Onassis? It was clear that they were there for some kind of summit meeting, and that there was a crisis of some sort. She did not imagine for a moment that it might concern her. The previous day, Onassis had confided that he was negotiating with a Middle East group to sell his share in the Olympic Tower on Fifth Avenue—"I *always* need money," he had grumbled when she teased him about being short of pocket money. "I have a *very* high-maintenance wife." She suspected that their arrival was connected to that deal.

The first night, Gaillet was invited to join Onassis and the new arrivals for dinner at the house. But it was apparent to her that her presence made the visitors uncomfortable; that she was inhibiting conversation, and had entered a world of secrets and ambiguity as well as male chauvinism. It was hard for her to tell whether they were ganging up on Onassis or trying to protect him. Perhaps it was about money. Money was usually the issue, she thought. Sometimes the Greeks talked to Onassis in Greek, but even when they used English they spoke in conspirators' undertones that blatantly excluded her from the conversation. Gaillet could tell that it was more than a case of a couple of xenophobic Greeks with a bad attitude toward women. She could sense the anger and accusation in their tone and body language; the process was brutal and whatever was being discussed was serious, unpleasant, and hard for Onassis to handle.

Only Karr remained detached. "Whatever it was he did, you could tell he was an expert at it," Gaillet recalled. The kind of guy who comes

in under everybody's radar, Karr was a listener, a collector of information. "David could destroy people overnight with the things he knew,"
Ronnie Driver would tell me later.[14] Although Gaillet could not recall a
single word he'd said all evening, "You just *felt* his presence, the way you
feel a chill when somebody walks over your grave," she said. But they *all*
seemed to be scared men to her. "That was the impression I had: They
were scared and they were blaming Ari for whatever it was that made
them scared," she told me.

But their fear seemed to diminish Onassis, too, and that made her
angry.

"I'd seen the way even the most successful men, the most distinguished and glamorous men on earth, behaved around Ari in New York.
He had an almost mystical power over people, he was a giant, but those
two Greeks seemed to . . . *despise* him is the only word I can think of:
they *despised* him."

At dinner, Gaillet continued to try to figure out exactly what was
going on between the Greeks as they argued in their own language. "Karr
was kind of quiet; he was hard to read, but the two Greeks were definitely afraid of something, and they were definitely angry, and it was
clear that Ari wasn't winning the argument. He was not retaliating. He
was not equal to whatever it was . . . their accusations? Had he stolen
money from them? Had he left them out of a big deal? Were they Greek
Mafia? I don't know. It was terrible to see Ari so crushed. He had aged
terribly. It was like being with a totally different person from the day
before when we made love on the beach."

Eventually, she'd seen enough and left. Only Onassis and Meyer
stood up and bid her goodnight.

The next day Gaillet remained on the yacht. In the evening, while Onassis dined at the villa with the Greeks and Karr, Johnny Meyer joined her
for dinner on the *Christina*. Married at least three times, Meyer really
preferred the bachelor life, and lived to perfect ways to make himself
indispensable to men like Onassis. An old Hollywood *spinmeister,* it had
been his special gift to be able to articulate, often colorfully, what actors
and producers felt but could not put into words. On the surface Meyer

was a Runyonesque character, but his drinker's face, mischievous eyes, and bulbous nose veiled a sharp and inquiring mind. He always remembered the indiscretions of others, and was a valuable source of information as well as prurient gossip for Onassis. He was a born storyteller and wonderful company.

Nevertheless, it was clear to Gaillet that he was not dining with her simply to keep her amused, or because he was bored with the company of the men at the villa. He wanted to find out if she was having an affair with Onassis. "He said things like, 'You two looked as if you had a secret you couldn't wait to talk about last night.'" She denied that she was sleeping with Onassis, but he continued to harp on about their closeness. "He would say, 'What do you two guys find to talk about, all alone out here?'" When she returned from the ladies' room, she saw him furiously scribbling notes on a damask table napkin. "I didn't think I'd said anything *that* interesting," she said.

But if Meyer was keen to find out why she was on the island, and how much Onassis had told her, Gaillet was equally curious about his visit, and what had riled up the Greeks so much. "We both knew we were pumping each other. I was trying to get Johnny to talk about what he was doing there with the Greeks and this man Karr: why they had turned up on the island unannounced; why Ari was so distressed." Eventually, Meyer suggested a truce: he'd stop grilling her, if she stopped grilling him. "He said, 'Ari did something bad and the Greeks are upset at him. They're volatile people, they fly off the handle, but it's just business. End of story. Now let's enjoy ourselves.'" Hélène knew that nobody got as rich as Onassis without doing things that are not strictly on the level—"there are no innocents in politics and business, bad things are done all the time," she told me—and she was happy to stop playing cat and mouse with Meyer.

They talked about other things. She was touched by the affection and warmth in Meyer's voice when he told her of the blows Onassis had suffered in the past couple of years. Everything that could go wrong had gone wrong in his life: his marriage, his airline, his investments were all in decline; the tanker business, still the linchpin of his fortune, was in depression. But nothing had hit Onassis harder than the death of Alexander and his conviction that his enemies had murdered the son to punish the father. It was all very Greek, said Meyer. But it was painful to watch a man who had come from poverty to such great wealth slowly disintegrate before his eyes. He had seen the same thing happen to his old boss, the eccentric billionaire Howard Hughes. But Onassis's decline was more private and far more poignant than Hughes's headlined disintegration.

His mind shattered, he often talked to his dead son, saw people who were not there, said disturbing things. "Ari says things that could cause a lot of trouble if anyone repeated them," Meyer told Gaillet. Aware that grief can affect people in different ways, she didn't at first pay much attention to this remark. But when Meyer asked whether Onassis had said anything that troubled her, she knew that he was fishing again. She was insulted not by the question itself but by the implication that she might repeat anything Onassis had told her privately.

She told me, "Men like Ari and Felix and André Meyer and Jimmy [Sir James] Goldsmith, all those really powerful men, when they connect together, when they get carried away with business, the women around them cease to exist. They become oblivious to your presence because their world revolves around mutual trust and confidence in one another's integrity, and once you have won their trust, proved your loyalty, they are able to completely relax. Bella Meyer [wife of Jackie's financial guru, André Meyer], Madeleine Malraux [wife of André Malraux, President Charles de Gaulle's closest adviser], we heard a lot of secrets because we were trusted; we acquired a lot of information dining at André Meyer's apartment at the Carlyle [Hotel in New York City]. I prided myself on my discretion. To be anything else would have been very stupid. And I was never stupid."

At the end of the evening, Meyer said he was flying back to Paris the following day and pressed Gaillet to go with him. As she recalled the conversation, he said: "The Greeks don't trust you, Hélène. They want you out of here." She told him, "If Ari wants me to leave, I'll leave. But I'd rather he told me himself. But he's not going to do that, Johnny, because Ari knows he can trust me completely."

Ari knows he can trust me completely . . .

If she had thought very hard indeed, Hélène Gaillet could not have said anything more calculated to disturb Johnny Meyer's peace of mind than those seven words.

Four years later, in a bar called *Ta-boo* on Worth Avenue in Palm Beach, Meyer would tell Brian Wells his version of how, shortly after Onassis had discharged himself from the hospital in New York in 1974, he "smuggled this great broad" Hélène Gaillet onto Skorpios. "Ari had a weakness for pedigree dames and Gaillet was running up his temperature, which scared the moussaka out of the Greeks." Onassis, he said, "was dying by inches," had a "troubled mind," and "the Greeks didn't want anybody getting too close to him in case he got it into his head to make a deathbed confession." [15]

But even when he was drunk, Meyer was skilled in the art of calibrated revelations, and refused for the moment to say what that "deathbed confession" might be, or why Onassis's mind was so troubled.

But if Meyer knew how to tantalize, Wells knew how to be patient. An experienced, well-connected, very tough former London reporter, no one had a better instinct for news or for spotting a story in the smallest fragment of information. Now executive editor of the *National Enquirer,* the biggest weekly tabloid in America, he knew that Meyer had spent his life saving the backsides of men like Errol Flynn, Howard Hughes, and Onassis. "He had lied for them, pimped for them, fixed hotel suites and alibis for them. When he told me he had a story about Onassis, Jackie, and Bobby Kennedy that was worth 'a million dollars minimum,'* I didn't doubt him for a second," Wells told me.

Well's oceanfront home on Ocean Ridge was only a couple of miles up the coast from Meyer's condominium apartment in South Palm Beach, and they had become good friends and regular drinking companions as well as proposed partners in Meyer's "million dollar memoirs." Meanwhile, Wells had started work on a treatment for the book. A reporter used to cutting to the heart of things and moving on, he became

---

* $2.8 million in today's money.

stymied by Meyer's procrastination over the revelation that he promised would "shock the world."

Wells knew that Christina continued to pay Meyer a generous retainer, and picked up his medical bills at the Mayo Clinic, where he was being treated for prostate cancer, and that he had also retained his Olympic Airways first-class travel card. These were perks not to be sniffed at. At the age of 72, he lived in a style that pleased him, a minor celebrity among people he liked, and he was surprisingly content. Even so, once he gave up his secret, Meyer's only problem would have been what to do with the money.

Wells told me, "It was a frustrating time. I knew Johnny was holding back a great story. I mean, a *great story.* I thought I only had to be patient and he would tell me everything. But I soon realized it wasn't going to be as simple as that. He wanted to tell me. He still talked with enthusiasm about the book, the waves it would make. But for years he had played a key role in keeping Onassis's nose clean, he was good at it, he was the best, and old habits die hard, I guess."

Nevertheless, Meyer had let hints drop, here and there; just enough for Wells to work out that the big revelation concerned a scandal involving Onassis and Jackie and Bobby Kennedy—and that it was still alarming enough to trouble Meyer three years after Onassis's death, and a whole decade after Bobby's assassination in Los Angeles.

Eventually, Wells told Meyer: "Look, Johnny, if this book is going to attract big publishing dollars, you've got to give me the good stuff."

Meyer said, "The good stuff is what gets you killed, kid."

The day after Gratsos, Georgakis, Karr, and Johnny Meyer left the island Onassis invited Gaillet to join him for dinner at the beach house. "I had a feeling I had to do something really special to warm up Ari's heart, because he was heartbroken; I could hear it in his voice," she would later recall. Although she traveled light, she had taken a simple Ossie Clarke chiffon evening dress to the island. "It was extremely dark green and brown, and cut on the bias with a belt that wrapped around the waist and tied at the back, which was completely bare. Two cuts across its long sleeves also revealed skin. It was incredibly sexy, and I never wore anything under it. I knew it would cheer him up."

When he saw how glamorous she looked, Onassis changed into a white dinner jacket and black tie. It was, she remembered, the nicest compliment he could have paid her. Nevertheless, the tux merely emphasized how drained he was, and how much the last twenty-four hours had taken out of him.

He apologized to her for the way she had been treated by his friends. "He said, 'Hélène, they were pissed at *me*. They didn't think I should be entertaining you here without a chaperone.' "

Gaillet understood their concern. She knew that he was planning to divorce Jackie, Christina had made that perfectly clear, and it wasn't hard to imagine how bad it would look, the trouble Jackie could make, and the millions it could add to her settlement, if she were to find out about Hélène's presence, alone with her husband, on the island. It was sure to arouse suspicions of promiscuity, and Gaillet—very much in love with Felix Rohatyn—did not want to get embroiled in a messy divorce. Even so, it had struck her as strange that it required "four very sharp guys" to point out these dangers to "one even sharper guy."

They dined at a table set up on the terrace, the lanterns glinting on the dark lenses of Onassis's glasses. Waiters in white gloves served Dom Perignon and beluga caviar. "It made you feel completely unmoored from reality," said Gaillet. Even for people to whom wealth had lost its novelty, this was another level of luxury. She told Onassis how much she liked the way the perfume of the eucalyptus trees mingled with the smell of salt from the sea. He said that eucalyptus came from the Greek, meaning well covered. "I always feel well covered here. This is where I come to lick my wounds," he told her. He had planted many of the trees on the island himself, all the trees and shrubs of the Bible; the eucalyptus was his favorite. They talked about religion, in which Gaillet found solace and inspiration. She had been raised a Catholic, as Jackie had, and he pressed her about her faith.

He was, she said, very interested in confession as a religious act, or perhaps more accurately in the idea of absolution. "He wanted to know whether a priest would forgive him his sins if he went to confession. The idea of acknowledging his sins, saying ten Our Fathers and five Hail Marys, and coming out all brand new I think really appealed to him. He was inquisitive about *everything*. And he could always draw people out. He loved me telling him stories about my life: I'd been rich; I'd been poor. I'd been a countess*; I'd been homeless. Nothing was too intimate or too personal for us to talk about."

Wickedly, he told her that his former brother-in-law and rival Stavros Niarchos kept two lovers for his mistress, a man and a woman, and liked to watch the three of them making love together; and Greta Garbo (a former lover, he claimed) found the smell of Montecristo cigars such a turn-on that she always traveled with her own supply in order to be able to offer one to a prospective lover whenever she wanted to put herself in the mood for love. "Hemingway got it right," he told her. "Women either break your heart or give you the clap." Aware that he was phobic about the Kennedys—"and that could get pretty tedious if you didn't get him off the subject fast"—she asked whether he also agreed with Hemingway's remark that the only difference between the rich and other people is the rich have more money? "He said, 'The rich get laid more, I know that . . . even that little runt Bobby Kennedy got laid more.' "

Onassis's hatred for Bobby was intoxicating, said Gaillet. "He and Bobby had a history going back long before Jackie, but Bobby had been dead six years and Ari *still* hated him . . . it was as if Bobby was still a threat to him. It was the strangest thing."

Sometime around two in the morning, Onassis walked her down to the beach, where a sailor waited in a launch to take her back to the *Christina*. They had gotten really close that night. Hélène had told him things she hadn't told another living soul, not even her priest. He'd told her things she was sure were not easy for him to talk about either. She

---

* She had been married at eighteen to Charles de Barcza, a Hungarian count, whose family had dominated the sugar refinery and distribution business in pre-war Hungary. The couple had two daughters. They separated after six years; de Barcza died before they were divorced, and technically Hélène had remained a countess.

got the feeling that he did not want her to go. He stood looking at the ocean, the lights of his yacht glittering across the dark water. Because his back was turned to her, it took Gaillet several moments to realize that he was talking: "like somebody praying, really . . . that is the only way I can describe it." As she strained to catch his words, he turned around and said matter-of-factly, "*You know, Hélène, I put up the money for Bobby Kennedy's murder.*"

Said as a simple statement of fact, the words conveyed no consciousness of wrongdoing, no angst or sense of guilt at all. Perhaps because Meyer had cunningly forewarned her of the possibility that Onassis might say something shocking, something that could cause a lot of trouble if anyone repeated it, she was not completely surprised. "I said something like, 'Bobby Kennedy? Oh, Ari.' He gave that little Levantine shrug of his, like *so what?*" Gaillet recalled the familiar gesture that conveyed his attitude toward life.

Part of her wanted to believe that Onassis's confession was a macho fantasy; a paranoid delusion that he had personally plotted Bobby Kennedy's death; she wanted to believe that was all it was. But as she had told Meyer a few nights earlier, she also believed that there were no innocents in politics and business—"bad things are done all the time." Moreover, she knew that Onassis was a man who would always take the maximum risk to get what he wanted.

"I was in a state of shock, of course, although I didn't realize it then," she later told me. Onassis had actually confessed his complicity in Bobby Kennedy's murder, and, in her heart of hearts, she feared that she believed him. "In fact, the more I thought about it, the more I became convinced that he was speaking the truth," she said. "It was part of Ari's charm that he would trust you with an extreme confidence. Perhaps it was his way of tying you closer to him."

Bobby's death on the night of his triumph in the California Democratic presidential primary election had removed the last obstacle to Onassis's marriage to Jackie, and life, she knew, is seldom that neat, or that convenient. It would take a separate book to describe the anguish she went through trying to decide what to do, or who she could or should tell, Gaillet would later explain her dilemma to me. It wasn't a question of conscience, or some moral principle; Bobby was dead, and Ari was dying; what she did or said now would not change that. So, she decided "never to speak of it to anyone, not even to Felix." And for thirty years it remained one of the secrets she planned to keep forever.

Why Jackie turned up on Skorpios in that late autumn of 1974 is still unclear. Perhaps she had heard of Gaillet's presence on the island and was simply following the instincts of a suspicious wife; perhaps Gratsos, a games player who knew how to use an opponent's self-interest, had suggested it. Certainly, remembering her conversation with Christina, Jackie was the last person Gaillet expected to see on the island at that moment. "But I was pleased she came. We always got along very well, and her arrival put my mind at rest," she said.

On her last night on Skorpios, Gaillet had dinner with Ari and Jackie. Whatever conflicts or problems there may have been between the most famous couple on earth, and whatever feelings Gaillet might have had about making love with Jackie's husband only a few days before," were kept to themselves as they dined under the stars that evening. But Jackie was always at her best in a crisis. "She could make peace with almost any problem," Gaillet said. "If she were angry at my being alone with her husband on the island, she never by as much as a raised eyebrow showed it."

Leaving the following morning before Onassis had woken, Gaillet slipped into his dressing room and wrote a message in lipstick on his mirror, expressing her gratitude and her love. She ended it with Onassis's favorite expression: Fate happens.*

She never saw him again, and four months later he was dead.

---

* A quarter of a century later, at her home on Long Island, Gaillet has written in lipstick on her dressing-room mirror an ironic parody of the fairy tale rhyme. Who is the fairest of them all? It concludes: Mirror, mirror on the wall / One day I will see it all / But by then it'll be too late / For I will have met my fate!

THIRTY

TOO MANY GHOSTS

By heaven, I'll make a ghost of him that lets me!

—WILLIAM SHAKESPEARE

On the morning of October 10, 1974, fifty-five days after her daughter's overdose, a maid found Tina dead in her room in the Hotel de Chanaleilles, the Niarchos mansion in Paris. First reports, quoting sources close to her husband, said that the death was caused by an overdose of barbiturates.[1] This was swiftly contradicted by a statement claiming that she had died of "a heart attack, or a lung edema." The conflicting explanations, and the fact that Niarchos had waited twenty-four hours before revealing the tragedy—and that the first person he had called after Tina's body had been discovered was not a doctor, but his lawyer in Switzerland[2]—prompted comparisons to the murky circumstances of Eugenie's death on Spetsopoula four years earlier.

Tina's death was more than a loss to Niarchos and Onassis: It was also a potential disaster. For when Tina told Christina about her father's complicity in Bobby Kennedy's murder, in order to silence her accusations of Niarchos's culpability in Eugenie's death, she had created a hostage to fortune for all of them. For although they were relatively safe while Tina was around to remind her daughter of how much was at stake, Christina's discretion was unnervingly unreliable when she was left unattended.

Within hours of arriving in Paris—and before her father could stop her—Christina had demanded and gotten a magistrate's warrant ordering a police autopsy on Tina's body. When they talked on the telephone a few evenings earlier, she said, her mother had told her that she had ordered a new autumn wardrobe, and was going to change her hairstyle.[3] Were these the plans of a woman who intended to take her own life?

If Niarchos was terrified of having his private life exposed to the scrutiny of yet another murder investigation—and in a country where he had far less clout than he had commanded in the colonels' Greece—Onassis was equally afraid of angering Niarchos. "Stavros is going to think I'm up to my ass in this with you," Onassis told his daughter when he caught up with her in Paris.[4]

That evening, he called Niarchos and assured him that he accepted that Tina's death had been a terrible accident; Niarchos had his total support. "We must make it clear that we both welcome the autopsy, and want to clear the air," Onassis told him. If he, Onassis, of all people, demonstrated his faith in his innocence, he said, "it will be harder for people to say that things are not all okay."[5]

Two days after the joint statement welcoming the autopsy was released to the press, the pathologists appointed by the public prosecutor's office concluded that Tina had died from a pulmonary edema, a swelling in the lung resulting from an excess of fluid, probably complicated by her barbiturate dependency. There were no traces of violence on the body, the *London Times* reported, once again evoking memories of Eugenie's fishy death.[6]

But despite their cynical pact, Niarchos remained nervous about Onassis. After the funeral in Lausanne—"My aunt, my brother, now my mother—what is happening to my family?" Christina had wept at the graveside—Niarchos returned to Paris and was not seen in public for months.

But it was fear, not grief, that made him remote. He spent twenty hours a day in bed, afraid to leave the house. His paranoia and heavy drinking combined with his lifelong hypochondria were an alarming mix. A Paris aide, who had previously worked for the British Foreign Office, and who was familiar with the psychosis of suspicion, recalls how

Niarchos had once sent her assistant to New York with his urine sample to be analyzed because he didn't trust the laboratories in Paris. "He was convinced that the French doctors were reporting back to Onassis," she says.[7]

At the beginning of November, under the name of Mr. Phillips, Onassis again checked into a New York hospital. He was going downhill rapidly, and Christina was spending more and more time with him (as Jackie was spending less and less). When his eyelids became too weak to stay open, Christina taped them up with strips of Band-Aid and, aware of his vanity, to hide the plaster, ordered darker lenses for his glasses. "This is God punishing you for all your sins," she told him.

"I never think about sin," he answered, his voice gruff with age and nicotine. "It's my nature."[8]

Although he discharged himself from the hospital after a few days, it was clear that he was now seriously ill. Before he returned to Athens, he had a further meeting with Roy Cohn. Over lunch at "21," he told the lawyer that he still wanted to divorce Jackie. Cohn recalled, "He wanted to know how much her silence was likely to cost him. I said, 'That's going to depend on what you want her to be silent about.' He said, 'Well, I've lived an interesting life.' I said, 'Then she'll make you pay for it.' He said, 'Yes, that's what I thought.' "[9]

Meanwhile, a further financial crisis grounded the entire Olympic Airways fleet. The Greek premier put in an emergency management team and declared the government's intention of repossessing the airline. Onassis flew to Athens in December apparently to negotiate the transfer. But each morning he confronted the government negotiators with a fresh list of demands and queries, denying points they had agreed on twenty-four hours before. "Ari's mind seemed to wander aimlessly, but perhaps even that was a ploy," said a government negotiator. "We didn't know whether we were dealing with a genius or with a man who was exactly what he seemed: a burnt-out case."[10]

*   *   *

"The Widow's a jinx, Papa," Christina told her father the day Jackie flew off for some skiing on the Swiss slopes of Crans-sur-Sierre, while he continued to fight to save his airline in Athens.[11] His daughter pressed him to keep the promise to get rid of Jackie that he had made to Alexander in Paris; she saw it as an unfulfilled obligation to her brother's memory, and on December 11, her twenty-fourth birthday, Onassis renewed the pledge—provided that Christina married Peter Goulandris.

Onassis loved to celebrate New Year's Eve, but 1974 felt unlike all the others, he told friends at the small dinner party he had arranged in Paris. Declining the champagne, he sipped watered whisky to soothe his bleeding gums. At the end of the evening, when he handed each guest a gift, he gave Costa Gratsos, his oldest friend in the world, his interest in their Haiti project.* It was the moment, Johnny Meyer said later, when Onassis acknowledged it was the last New Year's Eve he would ever see.

Would the final months of his extraordinary life have been any happier for him if there had been a successful conclusion of Omega? or the Haiti deal? or had he been able to enter into a peaceful, rewarding old age? Business had been his ruling passion all his life, he told Meyer near the end, and what comfort is there in it now?

He had seized everything from life except the prize of a happy ending.

On January 15, 1975, Onassis handed Olympic Airways back to the Greek government for sixty-nine million dollars. "To casual observers, it looked like a favorable settlement," wrote Lewis Beman in *Fortune* magazine.[12] "Actually, all of the money that Onassis was to get from the government was earmarked to pay Olympic's outstanding debts."†

---

* See Chapter 27, page 252.

† The final settlement left him with some $15 million, plus real estate worth roughly another $10 million. He was also allowed to sell two 707s to Jordan for $9 million, and to keep for his own use a Learjet and two helicopters, valued at $500,000. Adding it all up, reported Lewis Beman, even $35 million was not much of a return for the blood, sweat, and tears Onassis had poured into the business in the two decades he had fought to keep it going.

Two weeks later, on February 3, Onassis collapsed with severe ab-
dominal pains. His doctors diagnosed gallstones and stressed that because of his impaired nutrition, caused in part by his increasing difficulty
in chewing food, his failure to eat regularly during the Olympic takeover
battle (his weight had dropped forty pounds in eight weeks), and a bout
of flu, he was in an extremely vulnerable state. Jackie flew back from
New York; Christina returned from Gstaad, where she had been skiing
with Goulandris, whom she knew she would never marry. But since spe-
cialists summoned from Paris and New York could not agree about the
best course of action (a French liver specialist wanted to operate at once
to remove Onassis's gallbladder; a New York heart specialist believed he
was too weak for invasive surgery), Onassis himself made the critical
decision to return to Paris and have his gallbladder removed at the American Hospital in Neuilly.

Accompanied by Jackie and Christina, he arrived in Paris on the
evening of February 6. A mild east wind barely moved the crisp tangle of
ivy and creeper that covered the tall wrought-iron railings outside 88
avenue Foch.[13] "I want to walk from this car under my own steam. I
don't want those sons of bitches to see me being held up by a couple of
women," he said as the night lit up with flash and klieg.[14] He could walk
only with great pain, but he climbed the steps unaided and went inside.

Later that evening Johnny Meyer was having a drink at the bar of the
Georges V Hotel with a journalist from *Paris Match* when he was called
to the telephone. The caller's voice was faint, the voice of a very old person; Meyer could not even be sure whether it belonged to a man or a
woman. "I can't hear you," he said. "Who is this?" Who the hell did he
think it was? the caller demanded with an adrenalin rush of anger. Meyer
did not know what hit him first, he later said: the fact that it was Aristotle Onassis on the line—"or the realization that I was talking to a dead
man."[15]

In his bedroom on the fifth floor of 88 avenue Foch, his mansion apartment in the best part of the best part of Paris, with the sound of the city
muted by the thick old walls and the armor-glazed windows, Onassis
had absorbed the first sixty milligrams of the slow-release pyridostigmine

capsule that would get him through the night, and which gave him the brief surge of energy into which he crammed as much business as he could handle.

It was a large room, and everything in it was large—the antique brass bed, the stone fireplace, the mahogany furniture, and the pictures on the walls—everything except Onassis. Cologned and freshly shaven, dressed in dark blue pajamas and a blue silk robe custom-made for him at Lanvin, he seemed to be a shrunken effigy of himself. It was crushing to look at him, impossible not to.

"He said, 'You never expected to see me looking like this, old friend?' " Meyer would always begin the story the same way, with the same impersonation of Onassis's faintly rasping voice. Although he was, in his own words, a bag of skin and bones, he seemed to have the trapped energy of a ghost. The part of his brain that had made the calculations and judged the risks for more than sixty years missed nothing. Even dying, Meyer said, Onassis exuded a sense of power. He knew what the outcome would be, but he didn't flinch. "Aristotle Socrates Onassis"—Meyer relished the resonance of the name—"Aristotle Socrates Onassis had all the courage in the world."

Jackie had failed to put her tasteful Bouvier touch on this room: and the red velvet drapes, gilt and gesso dressing screen, and fake Watteaus hung on the gray silk walls gave it the look of a boudoir in an old-fashioned Paris brothel. According to Christina, the reason why her father refused to let Jackie change the décor of their bedroom was to remind his wife, during their intimate moments together, of "what she really was." [16]

But now the ravaged old man seemed an incongruous occupant in such a place. The room was hot and smelled of medicine and scent. "It was Christina's scent, jasmine and patchouli, the kind of scent that's good on furs; Jackie's was flowery, a young girl's perfume, best for outdoors and country weekends," Meyer would say each time he told the story. [17] It was the kind of detail that made him such a fine raconteur.

Onassis asked whether the press were still waiting in the street. "You are always newsworthy, Ari," Meyer told him truthfully, although they both knew that the media death watch had begun. It was the last privilege of people like him to give the public a show at the end, Onassis said with a glint of his old hubris. His speech was slurred. He held his chin to support the weight of his head and to help him talk. "But what a chase, Johnny. What a chase it's been." [18]

Meyer might have changed a line here, added a detail there, to enliven the narrative—there was no clock on the bedside table as he claimed, just photographs of Alexander and Christina, a small crucifix, and a pocket calculator—"His whole life was there on that tiny table, from the cradle to the grave; a goddamn calculator and a crucifix!"—but Meyer's story of that evening had a ring of truth that never changed: Onassis saying: "I was just a Greek kid who knew how to do his sums," Meyer taking his hand, and telling him: "Who ever heard of anybody dying of droopy eyelids?" And Onassis answering, "I don't want to go on living with ghosts, Johnny. Too many ghosts." [19]

At 11:50 the following morning Onassis left in a blue Peugeot from the underground garage and headed for the American Hospital. It was a mild, bright sunny day.[20] The car sped onto the inner boulevard Périphérique at the place de Maréchal de-Lattre-de-Tassigny and turned right along the boulevard de l'Amiral Bruix to the Porte Maillot. At noon, as the throng of photographers and reporters at the main entrance was distracted by the arrival of Jackie and Christina, he entered the hospital through the adjacent chapel, known to interns as the "artists' exit," since it is also the route to the morgue.

On Sunday, February 9, the day Onassis's gallbladder was removed, Christina called David Karr and asked him to arrange a suite for her at the Plaza-Athenée. "I don't want to be in the same house as that woman," she told him bitterly, alluding to her stepmother.[21] But Jackie refused to be provoked. "Those Greek crones [Christina and her aunts] don't bother me. Remember, I cut my teeth on the Kennedy women!", she told Johnny Meyer, who respected her strength, but little else.[22]

Around this time, Meyer had dinner in a small bistro on the Quai des Grands-Augustins with an old friend, Hollywood PR veteran Ernie Anderson. Both men were drawn to the kind of celebrities who had more to hide than they wanted to reveal. Meyer told Anderson that

Onassis had only a few weeks to live. He wasn't afraid of dying, Meyer said, but he didn't want Jackie around when he drew his last breath. "He told me that Ari was convinced Jackie had *malocchio*—the evil eye," Anderson later recalled.[23]

According to Anderson, Onassis had arranged to give Meyer a hand signal—"because he's scared that he won't be able to talk at the end"— when he felt it was time to get rid of her.

"I couldn't buy the *malocchio* horseshit," Anderson told me. He believed that the real reason Onassis wanted Jackie out of the way was that he knew that his death would precipitate a fierce battle between her and his daughter for control of his fortune. Anderson said to Meyer, "It's just family business, isn't it, Johnny?"[24]

It was much darker than that, Meyer told him. "He said, 'Ernie, you have no idea how fucking Greek it is.' "[25]

On February 22, a *Paris Match* reporter talked to one of the surgeons who had operated on Onassis. "Our last ally for saving him is his pride. And that is the final unknown quantity," he said. Onassis was still on a respirator and remained connected to a kidney machine. He looks, said a paparazzi photographer who had snatched a picture of him through the hospital window, as if "his life is being sucked out of him."

A few days later Johnny Meyer saw Onassis for the last time. "They wouldn't let me in, so I bribed a nurse, who slipped me in at lunchtime when all the guards went together to eat. Onassis had tubes in his arms, and in his nose, and it looked like he had tubes in his head, too. *All he could do was give me a little wave with his hand,*" he wrote in the notes for his unfinished autobiography. It was the signal he had told Anderson he had been waiting for, although he does not admit this in his notes, which continue unconvincingly: "But the doctor said he was getting better, so Jackie and I got on the plane and flew home. She went skiing up in New England . . ."

On Saturday, March 15, 1975, it rained for twelve hours and nine minutes in Paris; it was the longest rainfall of the winter, and when it stopped Aristotle Socrates Onassis was dead.[26]

THIRTY-ONE

DIRTY MONEY

**Aristotle Onassis rescued me at a moment
When my life was engulfed with shadows. He
meant a lot to me. He brought me into a
world where one could find both happiness
and love. We lived through many beautiful
experiences together which cannot be forgotten,
and for which I will be eternally grateful.**

**—JACQUELINE BOUVIER KENNEDY ONASSIS,
1929–1994**

Many felt that Jackie's statement, issued on her arrival in France (accompanied by her mother, her children, and Edward Kennedy), was compelled by propriety rather than any sense of loss. "I remember those big dark glasses and her enigmatic smile as she came off the plane at Orly," recalls Peter Stephens, the Paris bureau chief of the *London Daily Mirror* at that time. "We were all a bit shocked by that smile."[1]

"How long did she spend with my father?" Christina asked when she was told that Jackie had gone directly to the hospital. Informed that her stepmother had prayed by Onassis's body in the hospital chapel for seven minutes, she said: "She couldn't spare him even ten minutes at the end."[2]

The first public indication of a serious split between the two women came a few days later on the journey to the fishing village of Nidri, from where a launch would carry Onassis's body to Skorpios. It was a cold, overcast day; church bells tolled, women in black shawls lined the streets as the cortege passed through each tiny village. Suddenly, the motorcade stopped. Christina left the limousine she was sharing with Jackie and Ted Kennedy and joined her aunts in the second car.

Kennedy, she later explained, had tried to talk to her about Jackie's settlement. Christina knew that he was not there to share her grief, and that they would want to discuss business sooner or later—"she's bringing her muscle to fight you over the money," Meyer had warned her when he heard that Kennedy would accompany Jackie to the funeral[3]—but Kennedy's impatience to get on with it, his apparent contempt for proto-col, infuriated Christina.

Meanwhile, on Skorpios the sky darkened as the coffin—cut from one of the island's walnut trees and bearing a brass plate inscribed sim-ply: "Aristotle Onassis: 1900–1975"*—was carried up the hill to the tiny chapel, past lines of employees holding lighted candles. As the body of Aristotle Socrates Onassis was lowered into a vault alongside Alexan-der's, it began to rain.

On April 18, the *New York Times* revealed that shortly before his death Onassis had instructed Roy Cohn to start divorce proceedings. "Several friends of the Onassis family have said that Mrs. Onassis wants more money," claimed the page one story by John Corry. Described as being "bitterly hostile" to her stepmother, Christina had herself instigated the leak to the *Times,* including the fact that Jackie's settlement had been the minimum her father could get away with under Greek law.

In fact, the woman who had come to Onassis with such a dowry of his-tory had been dismissed with a fixed annual income of $200,000, to be revoked—and his executors and heirs were instructed to fight her "through all possible legal means" [4]—if she challenged the will in any way.

---

* Contradictory to the end, the dates on his tombstone read: 1906–1975.

Jackie's ears burned with humiliation. "According to what I was told by very reliable sources," Roy Cohn later told me, "Jackie was calling up Christina in Monte Carlo after the story had been printed, threatening that unless Christina put out a statement saying that everything had been all lovey-dovey and wonderful between her father and Jackie, she was going to make no end of trouble over the estate—*and everything else.*"[5]

Costa Gratsos was alarmed by the vehemence of Jackie's response to the *Times* story. In spite of Onassis's threat that she would lose everything if she made any kind of fuss, it was clear that she didn't intend to go quietly—or on the cheap. Gratsos, aware of how much was at stake, pleaded with Christina to "at least talk to the woman."[6] But Christina refused.

As stories of the bitterness between Christina and Jackie became increasingly public, rumors of how much Jackie wanted also began to surface. Stories that she was holding out for $20 million (worth ten times that in today's currency), surprised even her most admiring friends. The suspicion began to grow that she "had something on her deceased husband, that she knew something—no one could say what—that she was not supposed to know."[7] Sam White of the *London Evening Standard* believed that it was the only explanation for Christina's eventual surrender to her demands. "The law was on Christina's side, she wouldn't have given 'the little *pied-noir,** as she liked to call her stepmother, a penny more than she had to. So why did she give her so much?" The answer, White believed, was plain: "Jackie knew far too much about Ari's misdeeds."[8]

On the evening of April 20, forty-eight hours after the *Times* story broke, Jackie again called Christina. The two women spoke together for fifteen minutes—an inordinately long time for them—after which Christina summoned Johnny Meyer to avenue Foch. He found her sitting behind her father's desk.

"Does this thing work, Johnny?" she asked, indicating the lie detector machine that Mitchell WerBell had given Onassis for introducing him to the military arms buyers in Athens. Her father had trusted it, Meyer told

---

* *Pied-noir* is a derogatory term for the mixed race lower-class in French Algeria.

her. "In that case," she said, hitting the arms of her chair with both fists, the way her father had done, "I've got a problem." [9]

On April 22, the *Times* carried an agency report from Paris headlined:

Miss Onassis Denies Her Father Planned Divorce

The stories were "totally untrue," she declared through her French attorney. Her relations with her stepmother were based on "friendship and respect" and there were "no financial or other disputes separating them," she insisted. Two days later the two women met on Skorpios for the Greek Orthodox service marking the fortieth day after Onassis's death, and broke sacramental bread together, a symbolic act to send the soul of the deceased to heaven.

But when, later that evening, Christina tried to pin her stepmother down about the settlement ("How much do you want, Jackie? Give me a number," Christina demanded, in her version of their showdown), Jackie was polite but evasive, telling her stepdaughter that it was the wrong time to talk about money! Jackie was too smart to be drawn into negotiating on Skorpios, says Hélène Gaillet de Neergaard, who had remained a friend of both women, and had watched their maneuvering with amusement: "Jackie wanted Christina on her own turf in New York." [10]

Nevertheless, Christina had plenty on her plate, and it clearly suited her to put off the moment when she would have to settle with her stepmother. As well as taking over the reins of her father's business, she had launched a lawsuit to annul her mother's marriage to Niarchos—on the grounds that it was unlawful for Tina to have wed her brother-in-law. A great deal of money was at stake; one inside estimate put Tina's personal fortune at $270 million, which would go to Christina if her mother's marriage was annulled, and to Niarchos if not. Few, however, gave Christina much chance against Niarchos—a man who had been her uncle, her stepfather, and, she alleged, her attempted seducer; "He's slept with my grandmother (Arietta Livanos), my mother, and my aunt. I think it must be a generational thing with him," she told friends, claiming that Niarchos had made a pass at her when she was sixteen.

Although her father's old lieutenants were spinning stories about how well she was running the company, the truth of the matter seems to be that Christina was a mess—and, according to friends, often heavily self-medicated. David Karr claims it was he who talked her out of suing Niarchos over her mother's money. "I said, 'Christina, go and see the man, don't let lawyers screw it up.' "[11] Whose side Karr was on at this point will remain one of the many mysteries of his life. But a few weeks later, Christina announced that she had withdrawn her suit against Niarchos and that there was now "complete harmony" between them.

The talks, however, had been anything but friendly. Christina found it impossible to trust the man she still believed had murdered her aunt, and driven her mother to suicide. Nor did Niarchos believe that an Onassis would ever be capable of acting honorably by him. "My father was not easily unsettled, but Christina unsettled him all the time," Constantine Niarchos told a friend.

Few realized, of course, that the negotiations were part of a much more complicated and deadly game being played with cards marked in blood. At least one law enforcement department in Greece was still curious about the part Niarchos had played in his wife's death on Spetsopoula. If it became known that he had used Onassis's complicity in one murder to stop Christina's allegations of his own culpability in another, they may have viewed him as a man who was not totally innocent.

In the event, it was Niarchos who blinked first. He handed back six of her mother's ships, all Tina's jewelry and paintings, and most of her money, believed to be as much as $250 million. " 'Do what you want, take all the money,' my father eventually told Christina," Constantine Niarchos would later recall of his father's capitulation. "He said, 'I don't want any of it!' "[12]

Meanwhile, Jackie was beginning to wonder whether it had been such a smart move to postpone settlement day with Christina on Skorpios after all. Her stepdaughter had failed to keep several appointments in New York and was avoiding her calls. Jackie told friends that she had "not been able to get a single cent out of Greece, not even my allowance. I've had to borrow money and sell stock just to pay my bills."[13] When she appealed to Artemis, her sister-in-law advised her to be patient: Christina was a child, and still mourning her father, she said.

In fact, having ditched Peter Goulandris, despite the promise she had made to her dying father that she would marry the shipping heir, and far from being overcome with grief, Christina had embarked on an affair with Alexander Andreadis, an heir to a reputed $500 million banking fortune. By the end of June, she still had not made contact, and Jackie's patience was exhausted. She called Georgakis and told him that she wanted a settlement by July 22, the day Christina was set to marry Andreadis. Although she did not make any overt threats, said Georgakis, "Jackie spoke like a woman who knew she had a strong hand, and was prepared to play it to the hilt."

Christina got on well with Georgakis, but she had never confided in him as her brother had done, and the first indication he had that Christina had reached at least a preliminary settlement with Jackie came on Christina's wedding day. Three months after her father's death, four weeks after meeting Andreadis over coffee at the Athens Hilton, Christina's rush to the altar had appalled her family, but clearly not Jackie. "I do so love that child . . . at last I can see happy days ahead for her," she gushed at the reception, where her presence was as unexpected as the nuptials themselves.[14]

Nevertheless, it would take a further twelve months of acrimonious wrangling to agree on the final settlement of $26 million—plus an annual income of $150,000; the pick of Onassis's third and fourth century B.C. marble heads and torsos, terracotta horses, and icons; and many other antiquities, books, pictures, and *objects d'art* that twenty years on would swell Sotheby's auction of her possessions to over $34.5 million.*

---

* Jackie did not get it all her own way, according to David Karr, who claimed that Christina had insisted that her stepmother keep her father's name: Even if she remarried, the name Onassis must remain a prefix. But why would Christina have wanted the Onassis name to live on in the woman she blamed for all the misfortune that had befallen her family since she came into their lives? One explanation, suggested Karr, was

It was an extraordinary haul, which one of Onassis's English friends described as "pillage, unashamed plunder." Nevertheless, her greed—and exceedingly good taste—did not surprise anyone who knew what she had gotten away with; what shocked them was that Christina had allowed it to happen. "I just got tired of arguing with her," Christina told friends. But it was a fairly lame excuse, and once again aroused speculation about what cards Jackie had held.

Although it is inconceivable that she had found out about Onassis's part in Bobby Kennedy's assassination, it is possible that she at least suspected how her husband had made his last killing trafficking drugs aboard his tankers. This would certainly account for his partner Gratsos's jumpiness about the time it took Christina to settle with her.

But did it also explain why Jackie had no qualms about holding out for so much?

And was it the reason why, when Christina wrote out the final check—and "before the ink was dry, Jackie reached over and pulled [it] out of her hand" [15]—Christina laughed in her face?

"It was dirty money anyway, Johnny, and I reckoned she'd earned every penny of it," she told Johnny Meyer, enigmatically. [16]

---

that when Tina divorced Onassis and reverted to her maiden name, Onassis had been so hurt that Christina wanted to ensure that it never happened again, even in death. But the truth is more likely that she didn't want Jackie ever to forget how much she owed her father. Meyer told Brian Wells that when Jackie insisted on the right to choose her own name, Christina told her, "If you want my father's fucking money, you keep his fucking name."

EPILOGUE

Not every end is the goal.

—FRIEDRICH WILHELM NIETZSCHE, 1844–1900

B y 1988, Onassis, David Karr,* Tina Niarchos, Johnny Meyer,† Costa Gratsos,‡ William Joseph Bryan, Jr.,§ and Mahmoud Hamshari, were all dead. Aside from Christina, who had been in some kind of clinical denial of the conspiracy since her father's death thirteen years earlier, Yannis Georgakis was the only one left, as far as he was aware, who knew the truth about the crucial role that Onassis had played in the plot to kill Bobby Kennedy.

---

* David Karr was found dead by his valet on the bedroom floor *(cont. on page 300)*

†Johnny Meyer died on October 17, 1978, in a freak automobile accident in Florida. Returning from dinner, he stopped his Lincoln to relieve himself by the roadside (suffering from prostate cancer, he had special dispensation from the Mayo Clinic to urinate by the rear wheel of his car when taken short), when a transmission defect caused the car to jump into reverse, killing him instantly.

‡Costa Gratsos died in New York of kidney failure on December 2, 1981.

§William Joseph Bryan, Jr., was found dead in his hotel room in Las Vegas on March 19, 1977.

At seventy-three, and prone to hypochondria and bouts of mild depression, Georgakis was increasingly concerned about Alzheimer's and all the other degenerative problems that can occur with the approach of early old age. "One way or another, I knew that I was unlikely to be there for Christina for too much longer," he would later tell me. "The fact that she refused to discuss the future with me was worrying, as well as irritating. She completely refused to acknowledge what had happened [RFK's assassination], and Ari's part in it. As far as I could tell, she had wiped the memory of it completely from her mind. But it *had* to be discussed, you see. It couldn't be ignored."

Georgakis's wish to confront Christina was not because he needed to share the burden of that dangerous knowledge, but because he continued to believe that the secret, even after twenty years, could still surface.

---

of his Paris apartment on the morning of July 7, 1979. Karr had a fractured larynx, and blood was found on his pillow; a police surgeon concluded that he had died instantly of a cardiovascular attack. Karr's widow Evia hurried back from New York and got a court order stopping the funeral; his body was sent to the Institut Medico-Legal for a forensic examination. Again it was concluded that he had died of a massive heart attack. Nevertheless, Evia pressed on with her charges of foul play, filing an action that resulted in the case being classified under French law as: "Homicide charges against persons unknown." Evia Karr, and Karr's business partner Ronnie Driver, still believe that Karr was murdered, probably by the PLO, to whom Karr (according to Israeli intelligence sources quoted in *Fortune* magazine on December 3, 1979) was selling Russian arms. A few weeks before his death, Karr contacted Leslie Linder, a former movie agent, whom Karr had known during his own days as a producer at MGM. Although Linder had become a successful film producer, and was out of the agency business, Karr wanted Linder to represent his proposed memoirs, in which Karr confided to Linder he would reveal that Onassis had been behind Bobby Kennedy's murder. "I said, 'Why hasn't anybody suspected Onassis's involvement before, David?'" Linder recalls his skepticism at the time. Karr told him, "Because Ari was smart; he insulated himself too well." Linder agreed to talk to Karr again, together with Oscar Beuselinck, a London lawyer. "The last time I saw David was at the Connaught [hotel in London]. I'd got back from New York just in time to say goodbye before he left for Moscow, where he was opening a big hotel. He was full of piss and vinegar. He said he had all the evidence for the Onassis story in Paris, and promised to call me and Oscar [Beuselinck] as soon as he got back from Moscow. Next thing I heard, he'd been found dead in Paris," Linder told me.

The disclosure of an embarrassing document that Georgakis did not know existed; the appearance of an unexpected witness; the discovery of the tapes that Alexander had secretly made of his father's raving and self-incriminating conversations, and put in a safe deposit box with Sirhan's notebooks at the Plaza-Athenée, to dissuade Onassis from harming Fiona*; even a slip of Georgakis's own tongue; there were so many ways that, as Onassis had often said, "fate happens."

There had, Georgakis knew, been a few close-run things.

At the time of William Joseph Bryan, Jr.'s, death in a hotel room in Las Vegas in 1977, for example, Bryan was being investigated by a couple of reporters pursuing the Manchurian Candidate theory first raised by Bob Kaiser in his 1970 book *R.F.K. Must Die!* Intrigued by a newspaper interview in which Kaiser had stated that "Sirhan was programmed to kill Bob Kennedy and was programmed to forget the fact of his programming,"[1] investigate reporters Bill Turner and John Christian had begun the first serious search for Sirhan's programmer: the "shadowy someone" first mentioned in Kaiser's book.

---

* In July 1975, eschewing the promise that she would marry shipping heir Peter Goulandris that she had made to her father on his deathbed four months before, Christina married Alexander Andreadis, the playboy son of a banker who was being investigated for alleged embezzlement. When David Karr asked what a woman who had everything wanted for a wedding gift, Christina suggested the contents of her late brother's deposit box at the Plaza-Athenée, where Karr had been given a directorship by Lord Forte, in gratitude for Karr's help in acquiring the prestigious hotel for the British magnate. Nevertheless, Christina's request was a tall order. For in spite of the search Johnny Meyer had conducted of Alexander's apartment in Monte Carlo after his death, no key, no box number, no clue to the pseudonym Alexander might have used, nor, indeed, any confirmation that he had actually used the Plaza-Athenée vault at all, as Christina believed, could be found. In the early hours of August 18, two armed men entered the Plaza-Athenée and while one held the night staff at gunpoint, the other systematically broke open the safe deposit boxes in the vault. Robbery squad detectives and agents from two branches of French intelligence were involved in the investigation; the only thing everyone agreed on was that the raiders must have had an inside accomplice. Although there is no evidence that the raid had anything to do with Karr's search for the perfect wedding present for Christina, Georgakis said that she never again spoke of the missing safe deposit box and its explosive contents. "It was as if its existence had been completely washed from her memory, along with Ari's complicity in Bobby's murder," Georgakis told me.

The more closely Turner and Christian looked at Bryan, the more they believed that he was their man. Bryan's frequent boasts that he had been a CIA consultant in the Agency's mind control programme MKULTRA resonated with their working hypothesis that Sirhan had been programmed by someone working for, or close to, U.S. intelligence. The reporters had the right idea, but they'd gotten onto the wrong track; chasing down intelligence and military-industrial connections who feared that Kennedy's antiwar policies would be bad for their business, the reporters' promising line eventually ran into a brick wall.

Georgakis read Turner and Christian's book[2] with growing unease; Christina, for whom so much was at stake, showed no interest in it at all. "I wanted to shake her," said Georgakis, who feared that the famous Onassis luck would eventually run out: "The cork is going to come out of the bottle one day," he fretted pessimistically.

Christina had to be ready for that crisis if it came. In the early winter of 1988, Georgakis began to press her harder to discuss the matter with him; it was important that they had a stratagem, a game plan in place should the truth ever leak out. Whether it would be a convincing denial, or a plausible explanation of how Onassis had become embroiled in the tragedy, Christina's response had to be prepared, rehearsed, fine-tuned. But she still did not return his phone calls; she ignored his letters asking for a meeting. Georgakis thought that being told of her father's complicity in the assassination when Christina was "only a child" [actually she had been twenty-one, and in the throes of her first divorce] was at the root of her denial, and many of her other problems, too. "I think she resented having a secret she wanted to forget," Georgakis said.

Whether that was true or not, or whether Georgakis's theory had any psychological validity, Christina's disturbed state of mind, her addiction to amphetamines, and other pharmaceuticals, and her general physical deterioration during the past decade, were undoubtedly following her mother's fatal arc. Christina's life, Georgakis said, was a mess.*

---

* A revealing description of Christina's physical and mental state in the eighties, by her former stepbrother the Marquess of Blandford (Jamie), appeared in *Heiress,* a biography of Christina by former *London Daily Mail* columnist Nigel Dempster. "She was a night-

She had been married and divorced three more times, had had count-less, disastrous, and costly affairs, since her father's death; she handed one impecunious young lover who lived in Marseilles $10,000 in bills simply to pay for his phone calls to her in Switzerland. Georgakis's reaction was one of pure fury when he heard about this. "She is out of her mind," he told her aunt Artemis. Artemis agreed. Indeed, during one alarming episode in 1978, when Christina married former KGB agent Sergei Danyelovich Kauzov—a marriage that was set up by the egregious David Karr—Artemis wanted her forcibly examined by psychiatrists to prove that she was legally out of her mind.*

But Christina was always planning a new life for herself, and in the spring of 1988, she embarked on one of her occasional periods of reha-bilitation and self-improvement. "I'm feeling good today," she told Billy Keating, an art dealer who was also one of Jackie's closest friends in Lon-don. "I shall call Jackie and tell her I want my money back—or else!" She never explained what the "or else" might be; friends, however, were familiar with her complaint that her stepmother had robbed her blind over the settlement after Onassis's death, and that one day Jackie would get her comeuppance. "My hands were tied then; Jackie knew too much about Papa, but now I know plenty about her," she told another friend, adding: "Jackie refused to wear black when my father died, but my father was screwing her when she was still wearing her mourning outfits for Jack Kennedy. So what does that tell you about what widow's weeds mean to her?"

---

mare," said Blandford, recalling a visit to Christina's villa in St. Moritz. "She had to be entertained all day, all night. She was on a diet so no one else was allowed to eat, no one was allowed to get close to food, the servants saw to it! She could not sit still. We'd watch a video and half-way through—click!—just in the good part, she'd say. "Let's go to King's [the local nightclub]. Then just as we were enjoying ourselves at King's, she'd want to go back and watch the rest of the video. Then it was back again to King's, or the Dracula [another local club], five or six in the morning, just because of the pills. She was fat, her teeth were green, but nothing that a good toothbrush wouldn't sort out. She never actu-ally smelled, because [her maid] Eleni wouldn't let her smell, she was simply yucky."

* To help Kauzov get started on his own in the shipping business when the marriage ended in 1980, Christina gave him an 18,000-ton bulk carrier which she bought for $3.5 million; two months later she also bought him a 60,000 tanker for $4 million.

In May 1988, Christina and I met for lunch in Paris; it was the meeting at which I gave her the videotape of *The Richest Man in the World,* the miniseries of my biography of her father. Although I did not know it at the time, this was shortly after she met with Georgakis, where they finally got to discuss "the problem," as they called her father's part in Bobby Kennedy's assassination. The next time I saw Christina was in October. She had invited me to Paris to discuss my biography of Onassis that she had finally gotten around to reading ("I was afraid of what I might discover," she explained the delay); she also wanted to talk to me about something else that she said had been on her mind since we met in May.

She now realized what a dangerous thing her father had once done to protect his airline, she began, easing into the problem of the secret she had borne for so long. In 1968, she went on, Ari had paid an Arab terrorist protection money to keep his Olympic airline safe from skyjacking, and terrorist attack, which were a huge risk at that time. He later learned that the terrorist, a Palestinian named Mahmoud Hamshari, had used the money to finance the murder of Bobby Kennedy. Now she was afraid that if the story got out "in a bad way," people might believe that her father had paid to have Bobby Kennedy killed, to clear the way for his marriage to Jackie—a marriage that everybody knew Bobby opposed.

It was a good story; it was a wonderful story. The possibility that it might be spin and chaff, concocted to defuse any damaging revelations in the future, did not enter my mind and when Christina asked whether I would write a piece about it, I took the bait. We agreed to meet again, together with Georgakis, when Christina returned from a trip to Argentina to attend a friend's wedding.

The following month, Christina was found dead in a half-filled bath in her suite at the smart Tortugas Country Club in the suburbs of Buenos Aires. Rumors that she had commited suicide were rife.* Deeply

---

* Because her body had been moved and placed on a bed, and because the attending physician was denied access to the bathroom, where there might be evidence to show how she had died, he refused to sign the death certificate. The body was moved to a local clinic, and then to the Clinica del Sol in Buenos Aires, where it was examined by two physicians (who noted that her body weight at the time of death was 76 kg (167

shocked, Georgakis said he did not want me to go ahead with the story of the diverted ransom money. After he read *Ari* two years earlier, he told me that I had "missed the real story." I asked him whether the story Christina had told me was the story I had missed? It was part of it, but not *exactly,* he said enigmatically; he refused to discuss the matter further.

I could not have been more surprised when, at the beginning of 1993, just eleven months before he died, Yannis Georgakis told me that he was ready to tell me the "real story" of Aristotle Onassis—the story I had missed in *Ari.* I have no idea what arguments must have gone through his head—what debates he must have had with himself about loyalty and discretion; about conscience, and truth—in the four years since Christina's death, that brought him to that decision. Was it the desire of a conscientious man to put the record straight? Or a need to settle some old scores?

I must, he said, begin with the premise that, for Onassis, Bobby Kennedy was unfinished business from way back . . .

---

pounds). Once again, the issue of a death certificate was withheld pending an autopsy. This, together with the fact that a federal judge had been notified of "doubtful" death, fueled the rumors of suicide.

ACKNOWLEDGMENTS

Nemesis could not have been written without the cooperation and trust of many people over many years. Foremost among those people was Yannis Georgakis, whose tantalizing remark that my 1986 biography of Aristotle Onassis "missed the real story" was the genesis of this book, and my first thanks must go to him. His decision to tell me the "real story," was only the start of his contribution; his astute insights, and wise counsel given with no strings or conditions, made this book possible. "Nothing happens in isolation, look for the cause, find the connection," I repeated his advice like a mantra whenever a line of enquiry threatened to run into the sand after his death. My book would not have been diminished without his help, ideas, and contributions; it would simply not exist.

I am grateful, too, to Christina Onassis, whom I had known since her eighteenth birthday. Even after her father's death, when she withdrew her collaboration on *Ari,* I understood her apprehensions; and we remained friends. I ask her forgiveness for writing this book, and not the story she wanted me to write.

I am especially grateful to Hélène Gaillet de Neergaard, without whom my book would have been considerably less explicit, and infinitely less well informed about the last months of Onassis's life. Her decision to share with me the intimate details of her experiences on Skorpios provided the final piece of a jigsaw puzzle that otherwise might never have been satisfactorily finished. Her willingness to part with secrets that she had kept to herself for thirty years was an act of remarkable unselfishness. A woman of sensitivity, humor, and style, I am privileged to count her as a friend.

In the course of writing a book of this kind over so many years, one accumulates many debts, large and small; the range of people who came into my life and helped me in so many different ways is extensive. But none has been more generous than Brian Wells. His greatest contribution to this book—in addition to his Johnny Meyer anecdotes, and sources—was the optimism and friendship he unfailingly provided. I am profoundly thankful for both.

I regret that I can only briefly express my gratitude and thanks to many of those who follow, quite a few of whom gave me hours of their time and the benefit of years of experience: Bassam Abu Sharif, former Palestinian terrorist, hijacker, and Yasser Arafat's chief lieutenant, who provided crucial advice that helped me navigate the tricky waters of internecine intrigue of the Palestinian terrorist factions; Faisal Husseini, a founder member of Fatah; Uzi Mahnaimi, a former spymaster for a secret unit of Israeli military intelligence; Miltiadis Yiannakopoulos, Onassis's own spymaster and security adviser; Raymond Kendall, General Secretary of Interpol; Christopher Dobson, a former colleague who has covered terrorist activity on four continents, and generously shared with me the product of his own investigations into Mahmoud Hamshari, and the PLO; Evia Karr for her frankness, and trust; Eric Clark, a true authority on intelligence matters; Philip Melanson, Director of the Robert F. Kennedy Assassination Archives at Southeastern Massachusetts University, and author of the important *The Robert F. Kennedy Assassination: New revelations on the conspiracy and cover-up.*

Twenty years ago, the help and guidance Fiona Thyssen gave me on *Ari*, contributed immensely to the veracity of that book. When, in its early days, I confided to her the secret that is at the heart of this book, she said simply: "Why am I not surprised?" I will always be grateful for her friendship, encouragement, and trust.

I thank Beau Abbott, an invaluable source concerning Onassis's venture into the drug trafficking business; Robert Maheu, who talked with me at length about his role in destroying Onassis's Jiddah agreement, and of matters that helped me make sense of so much that would have remained inexplicable without his help; Bill Jordan, the night watch commander at Rampart detectives station on the night Sirhan was arrested; the late Miles Copeland, a CIA man whose experience had been principally in Arab countries, for important introductions in a world he knew so well; and Leslie Linder, whose encounter with David Karr helped me more than he knows.

Alfred C. Ulmer bravely gave me two long interviews shortly after suffering a calamitous stroke in 1995. Robbed of his ability to speak coherently, unable to write, his struggle to communicate with me was a truly Herculean effort; although frequently frustrated by his helplessness, he refused to be beaten. According to his CIA colleague Joseph Burkholder, Ulmer had been "an especially dynamic figure in the success of CIA's operations in Greece"; probably no intelligence agency in the world had tracked Onassis more assiduously or knew more about his private and business affairs than the CIA; and no CIA man had gotten closer to Onassis than Ulmer. The effort he made to answer my questions, direct me toward new leads, and connections, was one of the gutsiest performances I've ever witnessed. Without his bravery and determination, this book would have been less informed, and considerably less explicit. Sadly, he died in 2000. I offer my condolences and appreciation to his son, Alfred C. Ulmer III, who arranged those important interviews for me.

In Palm Beach, I received the utmost kindness and help from Emmet Whitlock; William Carter; Mrs. Horace Schmidlapp; Ernie Anderson; and Johnny Meyer's widow, Inez, who put up with my many questions with patience and charm.

I owe very special thanks to Joan Thring, who talked to me with humor and frankness about her flings with Onassis, and friendships with

Jackie and Lee. I am also grateful to Alastair Forbes; Linda S. Gordon, Rico Zermeno; Charles S. Hirsch; Anthony Montgue Browne; Ronnie Driver; Lord Forte; Jean Chalmers; Pierre Salinger; John and Paula Michel; Lilly Lawrence; Norma Quine; Richard de Combray; Gore Vidal; Alan Brien; Nigel Dempster; John Pearson; Lori Winchester; David Metcalfe, Geraldine Spreckles Fuller; Peter Stephens; Lady Carolyn Townshend; Suzanne Kloman; and Clementine De Thier.

And also to Peter Curnock, my lawyer and friend; and Jeanne Hunter, a constant source of encouragement and good advice, to whom I have countless debts of gratitude.

Since the origins of this book go back over two decades, there are many who were of much help are now sadly gone. For their contributions, trust and often their friendship, I remain indebted to Maria Callas; Tina Niarchos; Sir John Russell; Nigel Neilson; Roosevelt S. Zanders; Joseph Bolker; Sir Peter Bristow; Alan Campbell-Johnson; Charles Feldman; Roswell Gilpatric; Rupert Allan; Joe Fox; Roy Cohn; Costa Gratsos; Johnny Meyer; Yoko Tani; Margot Fonteyn; Artemis and Theodore Garofalides; Alexander Onassis; C.L Sulzberger; Spyros Skouras; Constantine Niarchos; Evelyn Lincoln; David Karr; Peter Payne; and Robin Douglas-Home, whose contribution to a profile of Lee Radziwill I wrote for *Cosmopolitan* magazine in 1968, became invaluable thirty years on.

Many fine journalists have generously shared their information and insights into the Kennedys with me, but none more so than Anthony Summers, the author of *Conspiracy,* on the assassination of John F. Kennedy; and *Goddess,* a biography of Marilyn Monroe, two lives that were curiously intertwined with that of Onassis. Summers and his writing partner, Robbyn Swan, opened up their files to me and let me borrow them by the boxful, helped with contacts, and entertained my wife and me at their home in Ireland.

Kitty Kelley, author of the seminal and perennially entertaining Jacqueline Kennedy biography *Jackie Oh!,* gave me guidance, reassurance, photographs, and priceless friendship. Her suggestions as to the most fruitful paths for further inquiry were always right on the money.

Two first-class researchers have helped me. In Washington, Jacqueline Williams never hesitated to burn the midnight oil on my behalf; without her tenacity and charm at least two important witnesses might not have been traced, or persuaded to talk to me. Aliki Roussin found many of her compatriots in Greece, London and Paris who had useful stories to tell, including Georgia Veta, Costa Haritakis, Yannis Paizis, Thanassis Yiannoukos, Lambis Costivis, Kata Kouzinow, and Helen Speronis. I am grateful to many fine Greek journalists and authors who went out of their way to answer my questions. I want particularly to thank Taki Theodoracopulos, Dimitri Rizos, Constantino Roussen, and the late newspaper publisher, Helen Vlachos. I am also indebted to Count and Countess Flamburiari, my friends, Milly and Spyro, for patiently explaining to me the mores and intricacies of Greek society.

I was fortunate in also having two peerless picture researchers on this project: Cathie Arrington and Suzanne Bosman. I thank them both. I also thank my agent, Andrew Lownie, for his care, and faith in this project from the moment I gave it to him.

I am lucky to be able to count Raymond Hawkey among my closest friends and oldest colleagues. From the start he kept the secret that is at the heart of this book. A fine writer, and distinguished designer, he never complained when he would put aside his own work to offer critiques of successive drafts of my manuscript. I hope the finished product justifies the time and concern he so ungrudgingly bestowed upon it.

At ReganBooks, I thank Calvert Morgan who saw the merits of the manuscript so quickly, and became a wise sounding board. I owe a great deal of debt to the amazingly calm and talented Aliza Fogelson, who edited it with skill, perception, and understanding. To Jessica Colter who assisted Aliza so well with deadlines loomed. Lawyer Christopher Goff who gave up his time to read the manuscript.

My biggest debt is to my family, who, for the past decade, in one way or another, have been caught up in my obsession with this story. Lisa Jane and William; Camilla and Clementine; Mark and Christine; even my brother, Michael; showed wonderful patience. Their love and unwavering trust were my sustaining strength.

Finally, this book would not have gotten this far without the understanding, and tireless support of my wife, Pamela, whom I love and admire more and more with each passing deadline.

NOTES

INTRODUCTION

1. Peter Evans, *Ari: The Life and Times of Aristotle Socrates Onassis*. New York: Summit Books, 1986.

CHAPTER ONE

1. Truman Capote, *Answered Prayers*. London: Hamish Hamilton, 1986.
2. Christopher Ogden, *Life of the Party: The Biography of Pamela Digby Churchill Hayward Harriman,* p. 257. London: Little, Brown & Co., 1994.
3. Author's interviews with Tina Livanos Onassis Blandford Niarchos.
4. Author's interviews with John (Johnny) W. Meyer.
5. Richard Rovere, *Senator Joe McCarthy,* p. 166. New York: Harper & Row, 1973.
6. William V. Shannon, *The Heir Apparent: Robert Kennedy and the Struggle for Power,* p. 55. New York: Macmillan, 1967.
7. Nicholas von Hoffman, *Citizen Cohn: The Life and Times of Roy Cohn,* p. 141. New York: Doubleday, 1988.
8. Arthur M. Schlesinger, *Robert Kennedy & His Times,* p. 112. London: Andre Deutsch, 1978.
9. Kirk LeMoyne "Lem" Billings, cited in Peter Collier and David Horowitz, *The Kennedys: An American Drama,* p. 202. New York: Summit Books, 1984.
10. George Skakel, cited in Jerry Oppenheimer, *The Other Mrs. Kennedy,* p. 179. New York: St. Martin's Press, pb ed., 1995.
11. Author's interviews with Aristotle Onassis.
12. Author's interviews with Sir John Russell.
13. Author's interviews with David Karr.

14. Author's interviews with Sir John Russell.

15. Capote, "La Côte Basque," *Esquire,* November 1975.

16. Author's interviews with Sir John Russell.

17. Schlesinger, p. 112.

18. Ibid., p. 108.

19. Shannon, p. 56.

20. *Boston Post,* April 12, 1953.

21. Author's interviews with Aristotle Onassis.

22. Author's interviews with Constantine Gratsos. This was a moot point, however. Onassis's rival and brother-in-law Stavros Niarchos was already unable to return to the United States because of a sealed U.S. Justice Department indictment against him for violations of the Ship Sales Act of 1946, which threatened his entire American operation. Since his own deals were identical to Niarchos's, they had even used some of the same people in Washington, it was clearly only a matter of time before the Justice Department caught up with Onassis.

23. From U.S. Office of Naval Intelligence report on Aristotle Onassis in response to FBI request made on 17 January 1943. Acquired by author under FOIA.

24. Author's interviews with Yannis Georgakis.

25. Nigel Hamilton, *JFK: Reckless Youth,* p. 494. New York: Random House, 1992. He was ordered to report to Naval Reserve Midshipmen's School Abbott Hall, Northwestern University, on July 22, 1942.

26. *Der Spiegel,* January 1953.

27. Willi Frischauer, *Onassis,* p. 151. London: Bodley Head, 1968.

28. L.J. Davis, *Onassis and Christina: The Amazing Story of a Fabulous Dynasty,* p. 62. London: Gollancz, 1987.

CHAPTER TWO

1. Joseph Burkholder Smith, *Portrait of a Cold Warrior,* p. 208. New York: G.P. Putnam's Sons, 1976.

2. Peter Grose, *Gentleman Spy: The Life of Allen Dulles,* p. 450. London: Andre Deutsch, 1995.

3. Author's interview with a confidential source.

4. Author's interview with Constantine Niarchos.

5. Inspector General, *CIA (IG) Report.* U.S. Government Printing Office, 1967.

6. Senate Select Committee on Government Operations with Respect to Intelligence Activities, *Alleged Assassination Plots Involving Foreign Leaders,* p. 77. U.S. Government Printing Office, 1975.

7. Author's interviews with Robert Maheu.

8. Ibid.

9. Leonard Mosley, *Dulles: A Biography of Eleanor, Allen, and John Foster Dulles and Their Family Network,* p. 243. New York: Dial Press, 1978.

10. Department of State documents obtained through FOIA.

11. July 1, 1954. CIA documents obtained through FOIA.

12. Author's interview with Robert Maheu.

13. Ibid.

14. Ibid.

15. Author's interviews with *London Evening Standard* Paris correspondent, Sam White, who had known Onassis since the early 1950s.

16. Arthur M. Schlesinger, *Robert Kennedy & His Times*, p. 151. London: Andre Deutsch, 1978.

17. Author's interviews with Brian Wells.

18. Robert Maheu and Richard Hack, *Next to Hughes*, p. 47. New York: HarperPaperbacks, 1992.

19. Ibid.

20. Author's interview with Alan Campbell-Johnson.

21. Author's interviews with Constantine Gratsos, cited in Peter Evans, *Ari: The Life and Times of Aristotle Socrates Onassis*, p. 121. New York: Summit Books, 1986.

22. Cited in Willi Frischauer, *Onassis*, p. 156. London: Bodley Head, 1968.

23. Spyridon Catapodis, affidavit sworn at the British consulate in Nice, September 27, 1954.

24. Ibid.

25. Author's interviews with Robert Maheu.

26. Department of State documents obtained through the FOIA.

27. Nicholas Fraser, Philip Jacobson, Mark Ottaway, and Lewis Chester, *Aristotle Onassis*, p. 153. Philadelphia: J.B. Lippincott, 1977.

28. Author's interviews with Aristotle Onassis. Evans, p. 143.

29. Fraser, Jacobson, Ottaway, and Chester, p. 155.

CHAPTER THREE

1. FBI director J. Edgar Hoover memorandum, September 16, 1953, FOIA.

2. Author's interviews with Rupert Allan, Grace Kelly's public relations guru and Monaco press adviser. In an interview with *Paris Match* (July 4, 1953), Onassis vehemently denied that he wanted to fly the Panamanian flag over the palace. He was content to leave the flag of Panama on his ships, he said. "Besides, there's a French-Monagasque tradition which limits the tonnage of the principality. Would I oppose that?" he asked.

3. Author's interviews with Aristotle Onassis.

4. Ibid.

5. Ibid.

6. Gardner Cowles, cited in Nicholas Fraser, Philip Jacobson, Mark Ottaway, and Lewis Chester, *Aristotle Onassis*, p. 191. Philadelphia: J.B. Lippincott, 1977.

7. Jack Kroll with Scott Sullivan, "Portrait of a Lady," *Newsweek*, September 27, 1982.

8. Author's interviews with Rupert Allan.

9. Kitty Kelley, *Jackie Oh!*, New York: Ballantine, 1979.

10. Gerald Clarke, *Capote: A Biography*, p. 271. London: Cardinal, 1989.

11. *The Pursuit of Happiness,* produced by Dore Schary.

12. "Kennedy was a Hollywood figure before he became President," remembers producer Milton Sperling. "He used to come out here as a senator to meet girls; he was out here looking for ladies and Charlie Feldman was finding them for him. There was no big secret." Ronald Brownstein, *The Power and the Glitter: The Hollywood–Washington Connection,* p. 147. New York: Pantheon, 1990.

13. Author's interview with Charles Feldman.

14. Gore Vidal, *Palimpsest: A Memoir,* p. 311. London: Andre Deutsch, 1995.

15. Charles Feldman was one of Monroe's closest friends in Hollywood and she often used his house on Coldwater Canyon for her trysts. "I saw (Marilyn) frequently at Charlie Feldman's house," her lover Elia Kazan wrote in his memoirs, "and we'd stay together until the morning light . . ." *Elia Kazan: A Life,* p. 408. London: Andre Deutsch, 1988.

16. Edward Klein, *All Too Human: The Love Story of Jack and Jackie Kennedy,* p. 195. New York: Pocket Books, 1996.

17. Clarke, p. 272.

18. Garry Wills, *The Kennedy Imprisonment,* pp. 23–24. New York: Pocket Books, 1981. Betty Spalding was astonished at Kennedy's intrusive questions: "He would say personal things to me. I mean ask me personal questions about women and marriage . . . and later he talked to me about his sex life with Jackie."

19. C. David Heymann, *A Woman Named Jackie: An Intimate Biography of Jacqueline Bouvier Kennedy Onassis,* p. 188. London: Mandarin, 1990.

20. Cited in Peter Collier and David Horowitz, *The Kennedys: An American Drama,* p. 209. New York: Summit Books, 1984.

21. Klein, p. 215.

22. Kelley, p. 54.

23. LeMoyne Billings, cited in Heymann, p. 192.

24. Hilton Als, "The Talk of the Town," *The New Yorker,* June 8, 1998, p. 32.

25. Collier and Horowitz, p. 209.

26. Klein, pp. 220–221.

CHAPTER FOUR

1. Author's interviews with Sir John Russell.

2. Author's interviews with Constantine Gratsos.

3. Author's interviews with Aristotle Onassis.

4. Peter Collier and David Horowitz, *The Kennedys: An American Drama,* p. 196. New York: Summit Books, 1984.

5. Author's interviews with Onassis.

6. Ibid.

7. Author's interviews with Hélène Gaillet de Neergaard.

8. Author's interviews with Constantine Gratsos. Also cited in Peter Evans, *Ari: The Life and Times of Aristotle Socrates Onassis,* p. 164. New York: Summit Books, 1986.

9. Author's interviews with Costantine Gratsos.

10. Sam White, "Monte Carlo Newsletter," *London Evening Standard,* April 10, 1959.

11. Ibid.

12. Author's interviews with Aristotle Onassis.

13. Author's interview with Lady Carolyn Townshend.

14. Tina Onassis to Max Beaverbrook, April 15, 1959. Beaverbrook Papers, House of Lords Records Office, London.

15. Beaverbrook to Tina Onassis, April 20, 1959. Beaverbrook Papers, House of Lords Records Office, London.

16. The Lord Beaverbrook Chair, Notre Dame University, endowed June 1956. Ronald Kessler, *The Sins of the Father: Joseph P. Kennedy and the Dynasty He Founded,* pp. 332–333. London: Hodder & Stoughton, 1996.

17. Anthony Montague Browne, *Long Sunset: Memoirs of Winston Churchill's Last Private Secretary,* p. 194. London: Cassell, 1995.

18. Author's interviews with Costa Gratsos, Yannis Georgakis, and others.

CHAPTER FIVE

1. Author's interview with Spyros Skouras.

2. Giovanni Battista Meneghini, *My Wife Maria Callas,* p. 285. London: Bodley Head, 1983.

3. Author's interviews with Aristotle Onassis. Cited in Peter Evans, *Ari: The Life and Times of Aristotle Socrates Onassis,* p. 55. New York: Summit Books, 1986.

4. Ibid.

5. Author's interviews with Johnny Meyer.

6. Franco Zeffirelli, *Zeffirelli,* p. 143. London: Weidenfeld & Nicolson, 1986.

7. Arianna Stassinopoulos, *Maria Callas,* p. 226. London: Hamlyn Paperback, 1981.

8. Cited in Meneghini, p. 288.

9. Anthony Montague Browne, *Long Sunset: Memoirs of Winston Churchill's Last Private Secretary,* p. 253. London: Cassell, 1995.

10. Meneghini, p. 289.

11. Montague Browne, p. 247.

12. Notebooks of Lady Sargant, the former Nonie Montague Browne.

13. Montague Browne, p. 255.

14. Meneghini, p. 290.

15. Ibid.

16. Cited in Montague Browne, p. 256.

17. Arianna Stassinopoulos, *Maria Callas,* p. 230. London: Hamlyn, 1981.

18. Montague Browne, p. 257.

19. Author's interviews with Artemis Garofalides.

20. Author's interviews with Tina Niarchos.

21. Nicholas Fraser, Philip Jacobson, Mark Ottaway, and Lewis Chester, *Aristotle Onassis,* p. 81. Philadelphia: J.B. Lippincott, 1977.

22. Author's interviews with Aristotle Onassis.

23. Meneghini, p. 302.

CHAPTER SIX

1. Author's interview with a source that does not wish to be named.

2. Giovanni Battista Meneghini, *My Wife Maria Callas.* London: Bodley Head, 1983.

3. Franco Zeffirelli, *Zeffirelli: The Autobiography of Franco Zeffirelli,* p. 209. London: Weidenfeld & Nicholson, 1986.

4. *London Daily Mail,* September 9, 1959.

5. Arianna Stassinopoulos, *Maria Callas,* p. 235. London: Hamlyn, 1981.

6. Author's interview with Spyros Skouras.

7. From a letter Tina wrote a London friend, who does not wish to be identified.

8. Author's interviews with Theodore Garofalides.

9. Johnny Meyer, interview with Brian Wells.

10. Cited in Peter Evans, *Ari: The Life and Times of Aristotle Socrates Onassis,* p. 93. New York: Summit Books, 1986; and author's interviews with Aristotle Onassis and Spyros Skouras.

11. Author's interviews with Constantine Gratsos.

12. Ibid.

13. Anthony Montague Browne, *Long Sunset: Memoirs of Winston Churchill's Last Private Secretary,* p. 257. London: Cassell, 1995.

14. Zeffirelli, p. 182.

15. Ibid.

16. Ibid., p. 210.

17. Ibid., p. 210.

18. Author's interviews with Aristotle Onassis.

19. Author's interviews with Yannis Georgakis.

20. Author's interviews with Aristotle Onassis.

21. Murray Kempton, cited in Arthur M. Schlesinger, *Robert Kennedy & His Times,* p. 205. London: Andre Deutsch, 1978.

22. Author's interview with Robin Douglas-Home.

23. Author's interview with David Metcalfe.

CHAPTER SEVEN

1. Author's interviews with Yannis Georgakis.

2. Carlo Curti, *Skouras: King of Fox Studios,* p. 274. Los Angeles: Holloway House, 1967.

3. Author's interviews with Rupert Allan.

4. Arthur Miller, *Timebends: A Life*, p. 403. New York: Grove Press, 1987.

5. Ibid., p. 399.

6. Author's interviews with Rupert Allan.

7. Ibid.

8. Ibid.

9. Peter Harry Brown and Patte B. Barham, *Marilyn: The Last Take*, p. 299. New York: Dutton, 1992.

10. Author's interview with Taki Theodoracopulos, the right-wing writer and Greek shipping heir. "Stas was bored, he didn't care about Lee, and was very fond of hookers. We used to go with hookers together," he told me.

11. Kitty Kelley, *Jackie Oh!*, p. 188. New York: Ballantine, 1979.

12. Drew Pearson, *Washington Post*, November 14, 1963.

13. Author's interviews with David Karr.

14. Gore Vidal, *Palimpsest: A Memoir*, p. 19. London: Andre Deutsch, 1995.

15. Presidential aide Dave Powers cited in DuBois, *In Her Sister's Shadow*, p. 147. London: Little, Brown & Co., 1995.

16. Gerald Clarke, *Capote: A Biography*, p. 384. London: Cardinal, 1989.

17. Author's interviews with Robin Douglas-Home.

18. Author's interviews with Yannis Georgakis.

19. Author's interview with a confidential source.

20. Lee Radziwill, "Why My Sister Married Aristotle Onassis," *Cosmopolitan*, September 1969.

CHAPTER EIGHT

1. Author's interview with Nadia Stancioff.

2. Nadia Stancioff, *Maria Callas Remembered*, p. 162. London: Sidgwick & Jackson, 1987.

3. Aliki Rousin's interview with Jeanne Herzog for the author.

4. Author's interviews with Aristotle Onassis.

5. Kitty Kelley, *Jackie Oh!*, p. 209. New York: Ballantine, 1979.

6. Author's interview with Evelyn Lincoln.

7. Ibid.

8. John H. Davis, *The Kennedy Clan: Dynasty and Disaster 1848–1984*, p. 414. London: Sidgwick & Jackson, 1985.

9. Author's interview with a confidential source.

10. Author's interviews with Onassis, Fiona Thyssen, Yannis Georgakis, and Costa Gratsos.

11. Author's interviews with Yannis Georgakis.

12. Carlo Curti, *Skouras: King of Fox Studios*, pp. 306–307. Los Angeles: Holloway House, 1967.

13. Author's interviews with Yannis Georgakis. Hoffa's claim that Bobby dressed as a girl

when he was young and that he had been a homosexual is also cited in Peter Collier and David Horowitz, *The Kennedys: An American Drama,* p. 222. New York: Summit Books, 1984.

14. Author's interview with Rupert Allan.

15. Anthony Summers, *Goddess: The Secret Lives of Marilyn Monroe.* London: Gollancz, 1985.

CHAPTER NINE

1. Peter Collier and David Horowitz, *The Roosevelts: An American Saga,* p. 478. London: Andre Deutsch, 1995.

2. Author's interview with Angier Biddle Duke, Kennedy's chief of protocol.

3. David Halberstam, *The Best and the Brightest,* p. 223. London: Barrie & Jenkins, 1972.

4. Collier and Horowitz, *The Roosevelts,* p. 466.

5. Roosevelt, cited in C. David Heymann, *A Woman Named Jackie: An Intimate Biography of Jacqueline Bouvier Kennedy Onassis,* p. 390. London: Mandarin, 1990.

6. Author's interview with a confidential source.

7. Heymann, p. 389.

8. DP (Dedichen Papers), month undated.

9. Ingeborg Dedichen, *Onassis, Mon Amour: Memoirs Collected by Henry Pessar,* p. 23. Paris: Pygmalion, 1975.

10. Ibid., p. 137.

11. Author's interviews with Aristotle Onassis.

12. Aristotle Onassis, London. DP, undated.

13. Ibid.

14. Ibid.

15. Cited in Richard J. Whalen, *The Founding Father: The Story of Joseph P. Kennedy: A Study in Power, Wealth and Family Ambition,* p. 300. London: Hutchinson, 1964.

16. Aristotle Onassis, New York. DP, August 11, 1940.

17. U.S. Office of Naval Intelligence Report on Aristotle Onassis in response to FBI request made on January 17, 1943. Acquired by author under FOIA.

18. Aristotle Onassis, New York. DP, September 4, 1940.

19. Ibid.

20. Dedichen to Onassis, Bagneres-de-Bigorre. DP, August 8, 1940.

21. Onassis to Dedichen. DP, August 23, 1940.

22. Dedichen to Onassis, Bagneres-de-Bigorre. DP, September 3, 1940.

23. Author's interview with Bill Carter.

24. Author's interview with Geraldine Spreckles.

25. Dedichen, p. 201.

CHAPTER TEN

1. "Mrs. Kennedy to Visit Greece for Two-Week Rest in October: President's Wife Will Stay with Sister Near Athens on Private Vacation," *New York Times,* September 18, 1963.

2. Author's interview with Johnny Meyer.

3. Author's interviews with Yannis Georgakis.

4. Ibid.

5. Ibid.

CHAPTER ELEVEN

1. Author's interview with Suzanne Kloman, the former Mrs. Franklin D. Roosevelt, Jr.

2. Author's interviews with Constantine Gratsos, Yannis Georgakis, and others.

3. Author's interviews with Constantine Gratsos.

4. From Suzanne Kloman's journal, by kind permission of the author.

5. Author's interview with Suzanne Kloman.

6. Author's interview with Theodore Garofalides.

7. Ibid.

8. John Pearson's interview with Irene Galitzine for the author.

9. Author's interview with Tina Livanos Onassis Blandford Niarchos.

10. Charles Spalding, quoted in Peter Collier and David Horowitz, *The Kennedys: An American Drama.* New York: Summit Books, 1984.

11. Kirk Lemoyne "Lem" Billings, quoted in Collier and Horowitz, *The Kennedys,* p. 196.

12. Author's interviews with Yannis Georgakis.

13. Nigel Hamilton, *JFK: Reckless Youth,* p. 68. New York: Random House, 1992.

14. Author's interviews with Artemis Garofalides and a source who wishes to remain anonymous.

15. Author's interviews with Artemis Garofalides.

16. Pearson interview with Irene Galitzine.

17. Author's interview with Richard Burton.

18. Author's interview with Evelyn Lincoln.

19. Author's interview with a confidential source.

20. Author's interview with Constantine Gratsos.

21. Ibid.

22. Gerald Clarke, *Capote: A Biography,* p. 383. London: Cardinal, 1988.

23. Author's interview with Robin Douglas-Home. One of Stas Radziwill's closest London friends, Douglas-Home originally told me this story for a profile I was writing on Lee Radziwill for *Cosmopolitan* magazine in 1967. It was eventually cut from the piece by the magazine's lawyers.

24. John Pearson's interview with Henry Brandon.

25. Author's interview with Roswell Gilpatric.

26. *Newsweek,* October 28, 1963.

27. Milton Viorst, "The Skeptics," *Esquire,* November 1968.

28. Author's interviews with Christina Onassis.

CHAPTER TWELVE

1. Author's interviews with Artemis Garofalides.

2. Author's interviews with Yannis Georgakis.

3. Alastair Forbes, "A Loss Unparalleled? Not Bloody Likely," *London Spectator,* September 28, 1996, pp. 45–46; author's interviews with Alastair Forbes.

4. Gerald Clarke, *Capote: A Biography,* p. 384. London: Cardinal, 1988.

5. Laurence Leamer, *The Kennedy Women: The Triumph and Tragedy of America's First Family,* p. 588. London: Bantam Press, 1994.

6. *Boston Globe,* October 29, 1963.

7. Peter Collier and David Horowitz, *The Kennedys: An American Drama,* p. 338. New York: Summit Books, 1984.

8. Nicholas Fraser, Philip Jacobson, Mark Ottaway, and Lewis Chester, *Aristotle Onassis,* p. 247. Also, "(The President) made it clear that Onassis would not be welcome to enter the United States until after the election": Leamer, p. 588.

9. Author's interviews with Johnny Meyer.

10. William Manchester, *The Death of a President,* p. 27. London: Michael Joseph, 1967. Commissioned by Jackie less than a month after the fact, it is the "official" version of John F. Kennedy's assassination.

11. Michael R. Beschloss, *Kennedy v. Khrushchev: The Crisis Years 1960–63,* p. 614. London: Faber & Faber, 1991.

12. C. David Heymann, *A Woman Named Jackie: An Intimate Biography of Jacqueline Bouvier Kennedy Onassis,* p. 393. London: Mandarin, 1990.

13. Manchester, p. 28.

14. Ibid.

15. Author's interview with Evelyn Lincoln.

16. Herbert S. Parmet, *JFK: The Presidency of John F. Kennedy,* p. 339. London: Penguin Books, 1984.

17. Mary Barelli Gallacher, ed. *Frances Spatz Leighton: My Life with Jacqueline Kennedy,* p. 267. London: Michael Joseph, 1970.

18. Author's interviews with Aristotle Onassis.

19. Author's interviews with Hélène Gaillet de Neergaard.

20. Author's interview with Maria Callas.

21. Author's interview with Evelyn Lincoln.

22. Author's interviews with Yannis Georgakis.

23. Author's interview with Costa Gratsos.

24. Manchester, p. 27.

25. Kitty Kelley, *Jackie Oh!,* p. 216. New York: Ballantine, 1979.

26. Evelyn Lincoln, *Secret Lives: Jackie.* Barraclough-Carey Productions for Channel Four and Discovery Communications, 1995.

27. Ibid.

28. Author's interview with Evelyn Lincoln.

29. Ibid.

30. Ibid.

31. Leamer, p. 586.

32. Author's interviews with Johnny Meyer.

33. Author's interview with Rupert Allan.

34. Author's interviews with Johnny Meyer.

35. Author's interviews with Yannis Georgakis.

36. Manchester, p. 28.

37. Benjamin Bradlee, *Conversations with Kennedy,* p. 219. New York: W.W. Norton, 1975.

CHAPTER THIRTEEN

1. "The Manchester Book: Despite Flaws & Errors, A Story that Is Larger than Life or Death," *Time,* April 7, 1967.

2. Author's interview with Aristotle Onassis.

3. Gerald Clarke, *Capote: A Biography,* p. 382. London: Cardinal, 1988.

4. Author's interview with Joe Fox.

5. William Manchester, *Death of a President,* pp. 613–614. London: Michael Joseph, 1967.

6. Gore Vidal, *Palimpsest: A Memoir,* p. 312. London: Andre Deutsch, 1995.

7. Barbara Kellerman, *All the President's Kin.* New York: Free Press, 1981.

8. Stanley Tretick, *Oral History Program,* p. 22. Boston: John F. Kennedy Library, 1966.

9. Manchester, p. 614.

10. Ibid.

11. Ibid., p. 603.

12. Author's interviews with Constantine Gratsos.

13. Author's interviews with Yannis Georgakis.

14. Robert D. Heini and Nancy G. Heini, *Written in Blood: The Story of the Haitian People, 1492–1971.* Boston: Houghton Mifflin, 1978.

15. Author's interviews with Yannis Georgakis.

16. Ibid.

17. Author's interviews with Constantine Gratsos.

18. Author's interviews with Yannis Georgakis.

19. Peter Collier and David Horowitz, *The Kennedys: An American Drama,* p. 315. New York: Summit Books, 1984.

20. Arthur M. Schlesinger, *Robert Kennedy & His Times,* pp. 665–666. London: Andre Deutsch, 1978.

21. Collier and Horowitz, p. 316.

22. Schlesinger, p. 666.

CHAPTER FOURTEEN

1. Edward Klein, *Just Jackie*. New York: Ballantine Books, 1998.
2. Author's interviews with Johnny Meyer.
3. Gore Vidal, *Palimpsest: A Memoir*, p. 311. London: Andre Deutsch, 1995.
4. Author's interviews with Yannis Georgakis.
5. Schlesinger, p. 666.
6. C. David Heymann, *A Woman Named Jackie: An Intimate Biography of Jacqueline Bouvier Kennedy Onassis*, pp. 479–480. London: Mandarin, 1990.
7. Richard Lee, "Ethel Kennedy Today," *Washingtonian*, June 1983.
8. Thomas C. Reeves, *A Question of Character: A Life of John F. Fitzgerald*, p. 236. Bloomsbury, London: 1991.
9. Author's interviews with Yannis Georgakis, Constantine Gratsos, and others.
10. Author's interview with Roy Cohn.
11. Ibid.
12. "Music," *Time*, March 26, 1965.
13. Author's interviews with Yannis Georgakis.
14. Garry Wills, *The Kennedy Imprisonment: A Meditation on Power*, p. 109. New York: Pocket Books, 1983.
15. Cited by Gore Vidal, *Book Week*, April 9, 1967.
16. Kitty Kelley, *Jackie Oh!*, p. 272. New York: Ballantine, 1979.
17. John Corry, *The Manchester Affair*, p. 163. New York: G.P. Putnam's Sons.
18. Author's interviews with David Karr.
19. Kelley, p. 273.
20. Jeff Shesol, *Mutual Contempt: Lyndon Johnson, Robert Kennedy, and the Feud That Defined a Decade*, p. 356. New York: Norton, 1997.
21. Roche to Lyndon B. Johnson. December 23, 1966. Cited in Shesol, p. 359.
22. Shesol, p. 356.
23. Author's interview with a confidential source.
24. Author's interview with Brian Wells.
25. Carl Blumay with Henry Edwards, *The Dark Side of Power: The Real Armand Hammer*, p. 148. New York: Simon & Schuster, 1992.
26. Author's interview with Ronnie Driver.
27. Author's interviews with Yannis Georgakis.
28. Milton Viorst, "The Skeptics," *Esquire*, November 1968.

CHAPTER FIFTEEN

1. Arthur M. Schlesinger, Jr., *Robert Kennedy & His Times*, p. 933. New York: Ballantine, 1978.
2. Garry Wills, *The Kennedy Imprisonment: A Meditation on Power*, p. 107. New York: Pocket Books, 1982.
3. Author's interviews with David Karr.

4. Drew Pearson, "The Washington Merry-Go-Round," *Washington Post,* March 3, 1967.

5. Author's interviews with Rupert Allan.

6. Author's interviews with Johnny Meyer.

7. Author's interviews with Alexander and Christina Onassis.

8. Author's interview with a confidential source at the American Hospital, Paris.

9. Peter Collier and David Horowitz, *The Kennedys: An American Drama,* p. 339. New York: Summit Books, 1984.

10. Author's interview with Sam White.

11. Priscilla Johnson McMillan, *Marina and Lee,* p. 220. London: Collins, 1978.

12. Author's interviews with Aristotle Onassis and Constantine Gratsos.

13. Anthony Montague Browne, *Long Sunset: Memoirs of Winston Churchill's Last Private Secretary,* p. 292. London: Cassell, 1995.

14. Appendix to Hearings before the Select Committee on Assassinations of the U.S. House of Representatives, Vol. XII, 1979.

15. De Mohrenschildt claimed that he had been regularly debriefed by the CIA after his geological surveys abroad: Warren Commission testimony, Vol. IX, p. 236.

16. Author's interviews with Yannis Georgakis and Brian Wells.

17. Senate Select Committee on Government Operations with Respect to Intelligence Activities, Alleged Assassination Plots Involving Foreign Leaders (henceforth Assassination Report), pp. 129–130 (U.S. Government Printing Office, 1975); author's interview with Robert Maheu.

18. Norman Mailer, *Oswald's Tale,* p. 458. New York: Random House, 1995.

19. Author's interviews with Yannis Georgakis.

20. Author's interview with Rico Zermeno.

21. Author's interviews with Yannis Georgakis.

CHAPTER SIXTEEN

1. All the details and dialogue in the story of Callas and Onassis's battle with Vergottis *(Onassis and Calogeropoulos v. Vergottis)* are based on court records (Lloyd's Law Reports: Queen's Bench Division Commercial Court before Mr. Justice Roskill: April 17–28, 1967) with background insights and opinions from three lawyers engaged in the case, all of whom asked that they not be named.

CHAPTER SEVENTEEN

1. Peter Evans, *Ari: The Life and Times of Aristotle Socrates Onassis,* p. 239. New York: Summit Books, 1986.

2. Author's interviews with Alexander Onassis.

3. Ibid.

4. Ibid.

5. Author's interviews with Fiona Thyssen-Bornemisza.

6. Ibid.

7. Author's interviews with Christina Onassis.

8. Author's interviews with Fiona Thyssen-Bornemisza.

9. Author's interviews with Yannis Georgakis.

10. Author's interview with Nigel Neilson.

11. Author's interviews with Constantine Gratsos.

12. C. M. Woodhouse, *The Rise and Fall of the Greek Colonels,* p. 8. London: Granada, 1985.

13. "Greece," *Time,* June 11, 1973.

14. Author's interviews with Constantine Gratsos.

15. Author's interviews with Sir John Russell.

16. Ibid.

17. Author's interview with a confidential source.

CHAPTER EIGHTEEN

1. Author's interviews with Johnny Meyer.

2. Author's interviews with Brian Wells.

3. Author's interviews with Johnny Meyer.

4. Author's interviews with Constantine Gratsos.

5. Author's interviews with Nigel Neilson.

6. Author's interviews with a confidential Fatah source.

7. Andrew Gowers and Tony Walker, *Behind the Myth: Yasser Arafat and the Palestinian Revolution,* p. 47. London: W.H. Allen, 1990.

8. Author's interviews with Mohammed Ibrahim.

9. Ibid.

10. Author's interview with Lord Forte.

11. Author's interviews with Yannis Georgakis.

12. Ibid.

13. Aliki Roussin's interviews with Miltiadis Yiannakopoulos.

14. Author's interviews with Yannis Georgakis.

15. Aristotle Onassis in undated letter from the Savoy Hotel, London, to Ingeborg Dedichen, c. 1936. Dedichen Papers.

16. Author's interviews with Yannis Georgakis.

CHAPTER NINETEEN

1. Author's interviews with Sir John Russell.

2. Author's interviews with Fiona Thyssen-Bornemisza.

3. Ibid.

4. Author's interviews with Nigel Neilson.

5. Author's interviews with Yannis Georgakis.

6. Aliki Roussin's interviews with Miltiadis Yiannakopoulos.

7. Author's interview with Alan Brien. Just down from Oxford, embarking on a jour-

nalistic career, he had been invited to stay at the Chateau de la Croe to write an article on Onassis for the now-extinct English magazine *Illustrated*. The assignment had been set up by Randolph Churchill in order to enable Brien to ghost a profile he himself had been commissioned to write for *Life*.

8. Aliki Roussin's interviews with Miltiadis Yiannakopoulos.

9. L. J. Davis, *Onassis and Christina: The Amazing Story of a Fabulous Dynasty*, p. 77. London: Gallancz, 1987.

10. Author's interviews with Yannis Georgakis.

11. Author's interviews with Fiona Thyssen-Bornemisza.

12. Author's interview with Johnny Meyer.

13. Author's interview with Lilly Lawrence.

14. Paul Mathias, cited in Kitty Kelley, *Jackie Oh!*, p. 291. New York: Ballantine, 1979.

15. Charles Bartlett Oral History, LBJ Library, Austin, Texas.

16. Terry Smith, "Bobby's Image," *Esquire*, April 1965.

17. LBJ Library, Austin, Texas.

18. Peter Evans, "Jackie: The New Woman," *Cosmopolitan*, September 1975.

19. Eugene McCarthy, *Up 'Till Now: A Memoir*, p. 196. San Diego: Harcourt Brace Jovanovich, 1987.

20. Fred Sparks, *The $20,000,000 Honeymoon*, p. 26. New York: Bernard Geis Associates, 1970.

21. Author's interviews with Yannis Georgakis.

22. Author's interviews with Brian Wells.

23. Author's interviews with Aristotle Onassis.

24. Author's interviews with Yannis Georgakis.

25. Ibid.

26. Arthur M. Schlesinger, *Robert Kennedy & His Times*, p. 921. London: Andre Deutsch, 1978.

CHAPTER TWENTY

1. Author's interviews with Ernie Anderson.

2. Author's interview with Johnny Meyer.

3. Ibid.

4. Author's interviews with Yannis Georgakis.

5. L.J. Davis, *Onassis and Christina: The Amazing Story of a Fabulous Dynasty*, p. 75. London: Gollancz, 1987.

6. Author's interviews with Yannis Georgakis.

7. Author's interviews with Sir John Russell.

8. Author's interviews with Yannis Georgakis.

9. Ibid.

10. Ibid.

11. Davis, p. 75.

12. Author's interview with Roosevelt S. Zanders.

13. Author's interview with Joan Thring.

14. Ibid.

CHAPTER TWENTY-ONE

1. Author's interviews with Rupert Allan.

2. Ibid.

3. Author's interviews with Joan Thring.

4. Jean Stein and George Plimpton, *American Journey: The Times of Robert Kennedy,* p. 293. London: Andre Deutsch, 1971.

5. Cited in Arthur M. Schlesinger, *Robert Kennedy & His Times,* p. 293. London: Andre Deutsch, 1978. Author's emphasis.

6. "A Life on the Way to Death," *Time,* June 14, 1968.

7. Stein and Plimpton, p. 293.

8. Bill Adler (ed.), *The Uncommon Wisdom of Jacqueline Kennedy Onassis: A Portrait in Her Own Words,* p. 138. New York: Citadel Press, 1994.

9. Author's interviews with Joan Thring.

10. Memorandum to J. Edgar Hoover: FBI documents, dated May 17, 1968, acquired through the FOIA.

11. Author's interviews with Joan Thring.

12. Ibid.

13. Ibid.

14. Schlesinger, p. 982.

15. Trial testimony, *The People of the State of California v. Sirhan Bishara Sirhan,* Superior Court, Crim. 1969. Case No. 14026, Lubic, Vol. 19, 5525.

16. The People's Exhibit 78, Autopsy Report.

17. Author's interviews with Ernie Anderson and Brian Wells.

18. Ibid.

19. Author's interviews with Joan Thring.

20. Ibid.

21. Ibid.

22. Author's interviews with Brian Wells and Ernie Anderson.

23. Author's interviews with Joan Thring.

CHAPTER TWENTY-TWO

1. Plimpton's June 5 interview, LAPD (audio tape).

2. Joseph Lahaiv, cited in William W. Turner and John G. Christian, *The Assassination of Robert F. Kennedy,* p. 197. New York: Random House, 1978.

3. Author's interviews with William C. Jordan.

4. LAPD Summary Report, p. 426.

5. Robert A. Houghton (with Theodore Taylor), *Special Unit Senator: The Investigation of the Assassination of Senator Robert F. Kennedy,* p. 89. New York: Random House, 1970.

6. Sirhan's background is extensively described in James W. Clarke's *American Assassins: The Darker Side of Politics,* pp. 79–85. Princeton, N.J.: Princeton University Press, 1982. It is also reported extensively in LAPD and FBI files, and documented in trial transcripts.

7. Trial testimony, Hashimeh, Vol. 16, pp. 4599–4607.

8. Clarke, p. 82.

9. Trial testimony, Sirhan, Vol. 17, pp. 4856, 4937.

10. Defendant's Exhibit D.

11. Defendant's Exhibits II and JJ.

12. Trial testimony, Sirhan, Vol. 17, p. 4898.

13. Ibid., p. 4937; trial testimony, Strathman, Vol. 19, pp. 5381–5406.

14. Trial testimony, Sirhan, Vol. 17, pp. 4924–4937.

15. Trial testimony, Lewis, Vol. 17, pp. 4787–4802; Defendant's Exhibit G.

16. Robert Blair Kaiser, *R.F.K. Must Die!,* p. 94. New York: E.P. Dutton, 1970.

17. Author's interviews with Mohammed Ibrahim.

18. Kaiser, 1970.

19. Patrick Seale, *Abu Nidal,* p. 42. London: Arrow Books, 1993.

20. Author's interviews with Mohammed Ibrahim.

21. Seale, p. 44.

22. Author's interviews with Brian Wells.

CHAPTER TWENTY-THREE

1. Kitty Kelley, *Jackie Oh!,* p. 263. New York: Ballantine, 1979.

2. Kelley, p. 263.

3. Author's interviews with Constantine Gratsos.

4. Author's interviews with Johnny Meyer.

5. Author's interviews with William Carter.

6. Author's interviews with Aristotle Onassis.

7. Author's interviews with Rico Zermeno.

8. Author's interview with Emmet Whitlock.

9. Author's interview with William Carter.

10. Author's interview with Billy Keating.

11. George Smathers, cited in Peter Collier and David Horowitz, *The Kennedys: An American Drama,* p. 367. New York: Summit Books, 1984.

12. Author's interviews with Aristotle Onassis.

13. Author's interviews with Joan Thring.

14. Willi Frischauer, *Jackie* London: Sphere, 1977.

15. Aliki Roussin's interviews with Constantine Haritakis.

16. "No Match for the Ari-Jackie Shaw," Taki Theodoracopulos, *London Sunday Times,* August 17, 1997.

17. Author's interviews with Nigel Neilson.

18. Ibid.

19. William H. Honan, *Ted Kennedy: Profile of a Survivor,* pp. 130–131. New York: Quadrangle Paperback, 1972.

20. Cary Reich, *Financier: The Biography of Andre Meyer,* p. 262. New York: Morrow & Co., 1983.

21. Author's interviews with Yannis Georgakis.

22. Author's interview with Pierre Salinger.

23. Author's interviews with Aristotle Onassis.

24. "I still Love Ari," *News of the World,* September 14, 1969.

25. Gerald Clarke, *Capote: A Biography,* p. 519. London: Cardinal, 1989.

26. Diana DuBois, *In Her Sister's Shadow,* p. 209. London: Little, Brown. 1995.

27. Aliki Roussin's interviews with Constantine Haritakis.

28. John Henry Cutler, *Cardinal Cushing of Boston,* p. 358. New York, 1970.

29. "The Kennedys: Identity Crisis," *Newsweek,* November 4, 1968.

30. Author's interviews with Yannis Georgakis.

31. Author's interview with Billy Keating.

32. Philip H. Melanson, *The Robert F. Kennedy Assassination: New Revelations on the Conspiracy and Cover-up,* p. 24. New York: Shapolsky Publishers, Inc., 1991.

33. John Lindsay, *Newsweek.* Cited in Stein and Plimpton, *American Journey,* p. 293; Newfield, *Kennedy,* p. 286; Schlesinger, *Robert Kennedy & His Times,* p. 968.

34. "Bomb Search Delays Mrs. Onassis' Flight," *New York Times,* December 20, 1968.

35. Author's interviews with Yannis Georgakis.

CHAPTER TWENTY-FOUR

1. Trial testimony, Nahas, Vol. 16, pp. 4575–4588; Hashimeh, Vol. 16, pp. 4591–4622; Mary Sirhan, Vol. 16, pp. 4671–4719.

2. Ibid., p. 440.

3. Trial testimony, Diamond, Vol. 24, p. 6998; author's emphasis.

4. Author's interviews with William Jordan.

5. Defendant's Exhibit J.

6. People's Exhibits 71, 72, and 73.

7. Kaiser, pp. 284–285.

8. Interview with a confidential source, identified by the author as Mohammed Ibrahim.

9. Trial testimony, Unruh, Vol. 12, pp. 3283, 3291; Sirhan, Vol. 18, p. 5217.

10. Author's interviews with William Jordan.

11. Trial testimony, Jordan, Vol. 16, p. 4448; trial transcripts, Sirhan tapes, Vol. 21, pp. 5971–6011. William W. Turner and John G. Christian, *The Assassination of Robert F. Kennedy,* p. 227. New York: Random House, 1978.

12. Turner and Christian, p. 227.

13. Philip H. Melanson, *The Robert F. Kennedy Assassination,* pp. 207–208. New York: Shapolsky Inc., 1991.

14. Kaiser, pp. 69–70.
15. Ibid.
16. Ibid., p. 536.
17. Kaiser, p. 147; author's emphasis.
18. Philip H. Melanson, *The Robert F. Kennedy Assassination: New Revelations on the Conspiracy and Cover-Up, 1968–1991.* New York: SPI Books, 1991.
19. F. Lee Bailey, with Harvey Aronson, *The Defense Never Rests,* pp. 159–161. New York: Stein & Day, 1971.
20. Peter Grose, *Gentleman Spy: The Life of Allen Dulles,* p. 393. London: Andre Deutsch, 1995.
21. Richard Helms in a letter seeking CIA chief Allen Dulles's approval for the MKULTRA project on April 4, 1953. Dulles gave his approval to the program on April 13, 1953. Neither Congress nor the president was informed about MKULTRA. Church Committee Report, Book I; Grose, pp. 392–393.
22. John Marks, *The Search for the Manchurian Candidate,* p. 183. London: Allen Lane, 1979.

CHAPTER TWENTY-FIVE

1. Author's interviews with Alan Campbell-Johnson.
2. Sally Bedell Smith, "The Secrets of Midas," *Vanity Fair,* August 1992; author's emphasis.
3. Author's interview with Yannis Georgakis. Preparing his unfinished autobiography with writer Brian Wells after Onassis's death in 1975, Johnny Myer, who was also present when Onassis played the Spetsapoula tapes, repeated this incident in very similar detail.
4. Author's interviews with Brian Wells.
5. Author's interviews with Yannis Georgakis.
6. Ibid.
7. Author's interview with Brian Wells.
8. Author's interviews with Joseph Bolker.
9. Author's interviews with Fiona Thyssen-Bornemisza.
10. Author's interviews with Joseph Bolker.
11. Ibid.
12. Ibid.
13. Ibid.
14. Ibid, first cited in Peter Evans, *Ari: The Life and Times of Aristotle Socrates Onassis,* p. 255. New York: Summit Books, 1986.
15. Author's interviews with Fiona Thyssen-Bornemisza.
16. Ibid.
17. Ibid.
18. Ibid.
19. Ibid.

20. Author's interviews with Joseph Bolker; author's emphasis.

21. *Eleftherotipia,* Athens, 26 November, 1975.

22. Author's interviews with Fiona Thyssen-Bornemisza.

23. Ibid.

24. Author's interviews with Yannis Georgakis.

CHAPTER TWENTY-SIX

1. Author's interview with a source who does not wish to be named.

2. Nicholas Fraser, Philip Jacobson, Mark Ottaway, and Lewis Chester, *Aristotle Onassis,* p. 319. Philadelphia: J.B. Lippincott, 1977.

3. Willi Frischauer, *Jackie.* London: Michael Joseph, 1976.

4. Fraser, Jacobson, Ottaway, and Chester, p. 319.

5. Author's interviews with Rico Zermeno.

6. Fraser, Jacobson, Ottaway, and Chester, p. 295.

7. Author's interviews with Fiona Thyssen-Bornemisza.

8. Ibid.

9. Author's interviews with Roy Cohn.

10. "Don't Mess with Roy Cohn: The Legal Executioner," Ken Auletta, *Esquire,* December 1978.

11. Author's interviews with Roy Cohn.

12. David B. Tinnin, with Dag Christensen, *The Hit Team,* p. 82. Boston: Little, Brown, & Co., 1976; George Jonas, *Vengeance,* p. 187. London: Collins, 1984.

13. Author's interviews with Yannis Georgakis.

14. Roy Rowan, "The Death of Dave Karr and Other Mysteries," *Fortune,* December 3, 1979.

15. Carl Blumay with Henry Edwards, *The Dark Side of Power: The Real Armand Hammer,* p. 313. New York: Simon & Schuster, 1992.

16. Author's interviews with Mohammed Ibrahim.

17. Andrew Gowers and Tony Walter, *Behind the Myth: Yasser Arafat and the Palestinian Revolution.* London: W.H. Allen, 1990.

18. Author's interview with a confidential source.

19. Author's interviews with Fiona Thyssen-Bornemisza.

20. Ibid.

21. Author's interview with Donald McGregor.

22. Author's interviews with Fiona Thyssen-Bornemisza.

23. Ibid.

24. Ibid.

25. Author's interviews with Constantine Gratsos.

26. Author's interviews with Fiona Thyssen-Bornemisza.

27. Aliki Roussin's interview with Costas Haritakis.

28. Ibid.

29. Author's interviews with Brian Wells.

30. Ibid.

31. Author's interviews with Fiona Thyssen-Bornemisza.

32. Aliki Roussin's interview with Georgia Veta.

CHAPTER TWENTY-SEVEN

1. Author's interview with Yannis Georgakis.

2. Nicholas Fraser, Philip Jacobson, Mark Ottaway, and Lewis Chester, *Aristotle Onassis,* p. 334. Philadelphia: J.B. Lippincott, 1977.

3. Ibid., p. 335.

4. Author's interviews with Fiona Thyssen-Bornemisza.

5. Fraser, Jacobson, Ottaway, and Chester, p. 330.

6. Author's interview with Constantine Konialidis.

7. Author's interview with Yannis Georgakis.

8. Author's interview with Sir John Russell.

9. Author's interviews with Artemis Garofalides.

10. Author's interviews with Yannis Georgakis; also Aliki Roussin's interviews with Miltiadis Yiannakopoulos.

11. Author's interview with Emmet Whitlock.

12. Author's interviews with Yannis Georgakis.

13. Author's interviews with Rico Zermeno.

14. Author's interviews with David Karr.

15. Author's interviews with Yannis Georgakis.

16. Author's interview with a confidential source.

17. Penny Lernoux, *In Banks We Trust: Bankers and Their Close Associates: The CIA, the Mafia, Drug Traders, Dictators, Politicians, and the Vatican,* p. 94. Garden City, N.Y.: Anchor Press/Doubleday, 1984.

18. Author's interview with a confidential source.

19. Jim Drinkhall, *Wall Street Journal,* April 18, 1980.

20. Ibid; also Alfred McCoy, *The Politics of Heroin in Southeast Asia,* pp. 130–144, New York: Harper & Row, 1972; Henrik Kruger, *The Great Heroin Coup,* pp. 15, 130–131. Boston: South End Press, 1980.

21. Jim Drinkhall, *Wall Street Journal,* April 16, 1980.

22. Lernoux, p. 79.

23. Jim Hougan, *Spooks: The Haunting of America—The Private Use of Secret Agents,* p. 11. New York; William Morrow & Co., 1978.

24. Author's interview with former DEA undercover agent, Basil "Beau" Abbott.

25. From the affidavit of William D. Coller, former supervisor and special agent of the DEA, sworn for "the purpose of explaining the working relationship Basil Abbott had with the United States Government, and particularly the . . . Drug Enforcement Administration." May 21, 1997.

26. Author's interviews with Basil Abbott.

27. Ibid.

28. Ibid.

29. Author's interview with a confidential source.

30. Coller affidavit, May 21, 1997.

31. Author's interviews with Basil Abbott.

32. The Office of Naval Intelligence report, chapter 1.

CHAPTER TWENTY-EIGHT

1. Author's interviews with Yannis Georgakis.

2. Erich Morgenthaler, *Wall Street Journal,* quoted in L.J. Davis, *Onassis and Christina: The Amazing Story of a Fabulous Dynasty,* p. 185. London: Gollancz, 1987.

3. Nigel Dempster, *Heiress: The Story of Christina Onassis,* p. 100. New York: Grove Press, 1989.

4. L.J. Davis, p. 132.

5. Rosemarie Kanzler, quoted by Sally Bedell Smith, "The Secrets Of Midas," *Vanity Fair,* August 1992.

6. Davis, p. 192.

CHAPTER TWENTY-NINE

1. Author's interviews with Artemis Garofalides, Christina Onassis, and Lady Carolyn Townshend.

2. Author's interviews with Hélène Gaillet.

3. Ibid.

4. Ibid.

5. Ibid.

6. Ibid.

7. Author's interviews with Yannis Georgakis.

8. Author's interview with Count Flamburiari.

9. Author's interview with Yannis Georgakis.

10. Author's interview with Joseph Bolker.

11. Author's interview with Yannis Georgakis.

12. Author's interviews with Brian Wells, Johnny Meyer's Palm Beach friend and proposed ghostwriter of his unfinished memoirs.

13. Author's interview with Ronnie Driver.

14. Ibid.

15. Author's interview with Brian Wells.

CHAPTER THIRTY

1. Bernard Valery, the *New York Daily News*'s Paris correspondent, later told the author that his source was in fact Niarchos himself.

2. L.J. Davis, *Onassis and Christina: The Amazing Story of a Fabulous Dynasty*, p. 193. London: Gollancz, 1987.

3. Rosemarie Kanzler, an old friend of the family, later confirmed this. "She had just bought seventeen dresses, had her hair done and her face beautifully made up. You don't do that when you don't want to live." Sally Bedell Smith, "The Secrets Of Midas," *Vanity Fair*, August 1992.

4. Author's interviews with Johnny Meyer.

5. Author's interviews with Yannis Georgakis.

6. "Lung Ailment Caused Death of Tina Niarchos," *London Times*, October 14, 1974.

7. Author's interview with a confidential source.

8. Author's interviews with Christina Onassis.

9. Author's interviews with Roy Cohn.

10. Author's interview with a confidential source.

11. Author's interviews with Constantine Gratsos.

12. Lewis Bernan, "The Reality Behind the Onassis Myth," *Fortune*, October 1975.

13. National Meteorological Institute, Paris.

14. Author's interview with Christina Onassis.

15. Author's interview with Johnny Meyer.

16. Author's interviews with Christina Onassis.

17. Author's interviews with Johnny Meyer.

18. Ibid.

19. Ibid.

20. National Meteorological Institute, Paris.

21. Author's interview with David Karr.

22. Author's interviews with Johnny Meyer.

23. Author's interview with Ernie Anderson.

24. Ibid.

25. Ibid.

26. National Meteorological Institute, Paris.

CHAPTER THIRTY-ONE

1. Author's interview with Peter Stephens of the *London Daily Mirror*.

2. Nigel Dempster, *Heiress: The Story of Christina Onassis*, p. 109. New York: Grave Press, 1989.

3. Author's interviews with Brian Wells.

4. David Pauly with Scott Sullivan, "I, Aristotle Onassis . . .", *Newsweek*, July 28, 1975.

5. Author's interview with Roy Cohn; first cited without emphasis in Peter Evans, *Ari: The Life and Times of Aristotle Socrates Onassis,* p. 310. New York: Summit Books, 1986.

6. Author's interviews with Yannis Georgakis.

7. L.J. Davis, *Onassis and Christina: The Amazing Story of a Fabulous Dynasty,* p. 217. London: Gollancz, 1987.

8. Author's interviews with Sam White.

9. Author's interviews with Brian Wells.

10. Author's interviews with Hélène Gaillet de Neergaard.

11. Author's interviews with David Karr.

12. Dempster, p. 122.

13. Davis, p. 209.

14. Steven V. Roberts, "Christine Onassis Is Unexpected Wed to a Greek Banking and Shipping Scion," *New York Times,* July 23, 1975.

15. Gore Vidal, *Palimpsest: A Memoir,* p. 310. London: Andre Deutsch, 1995.

16. Author's interviews with Brian Wells.

EPILOGUE

1. Philip H. Melanson, *The Robert F. Kennedy Assassination.* New York: SPI Books, 1991.

2. William W. Turner and Jonn G. Christian, *The Assassination of Robert F. Kennedy: A Searching Look at the Conspiracy and Cover-Up 1968–1978.* New York: Random House, 1978.

DATE DUE			